Spon's Practical Guide to Alterations and Extensions

Also available from Taylor & Francis

Spon's estimating cost guide to minor works
Alterations and repairs to fire, flood, gale and theft damage
3rd edition
B. Spain Pb: ISBN 978–0–415–38213–7

Spon's house improvement price book 3rd edition
B. Spain Pb: ISBN 978–0–415–37043–1

Understanding the CDM regulations
O. V. Griffiths Pb: ISBN 978–0–419–22420–4

Understanding the building regulations
S. Polley Pb: ISBN 978–0–415–45272–4

Understanding building failures
J. Douglas & B. Ransom Pb: ISBN 978–0–415–37083–7

Fundamental building technology
A. Charlett Hb: ISBN 978–0–415–38623–4
Pb: ISBN 978–0–415–38624–1

Spon's Practical Guide to Alterations and Extensions

Second Edition

Andrew R. Williams

Taylor & Francis
Taylor & Francis Group
LONDON AND NEW YORK

First edition published 1995 by Spon Press

Second edition published 2008
by Taylor & Francis
2 Park Square, Milton Park, Abingdon, Oxon OX14 4RN

Simultaneously published in the USA and Canada
by Taylor & Francis
270 Madison Ave, New York, NY 10016, USA

Taylor & Francis is an imprint of the Taylor & Francis Group, an informa business

Typeset in Plantin by
Integra Software Services Pvt. Ltd, Pondicherry, India
Printed and bound in Great Britain by
Antony Rowe Ltd., Chippenham, Wilts

British Library Cataloguing in Publication Data
A catalogue record for this book is available from the British Library

Library of Congress Cataloging in Publication Data
Williams, Andrew.
Spon's practical guide to alterations & extensions / Andrew Williams. -- 2nd ed.
p. cm.
Includes bibliographical references and index.
ISBN 978-0-415-43426-3 (pbk. : alk. paper) --
ISBN 978-0-203-93204-9 (ebook) 1. Buildings--Additions. I. Title.
TH4816.2.W55 2008
690′.837--dc22
2007031731

ISBN10: 0–415–43426–2 (pbk)
ISBN10: 0–203–93204–8 (ebk)

ISBN13: 978–0–415–43426–3 (pbk)
ISBN13: 978–0–203–93204–9 (ebk)

For
Geraldine, my wife
and
Thomas Grenville Williams, my late father
and
Colin Paul Williams, my late brother

About the author

Andrew R. Williams is the managing director of a surveying practice. The practice was first established in 1977 and since then has provided a wide variety of quantity surveying and building surveying services.

www.andrewrwilliams.co.uk

By the same author

Domestic Building Surveys, E&FN Spon, London. *A Practical Guide to Alterations and Extensions*, E&FN Spon, London. *A Practical Guide to Single Storey House Extensions*, Building Trade Magazine.

Contents

Preface to Second Edition

Regulations change with alarming regularity and once again I have set about updating my book. The reasons for writing it in the first place are the same as given in the first edition.

Andrew R. Williams FRICS, FCIOB, FBEng, DipHi
Andrew R. Williams & Associates Limited
Building Services (Technical) Limited
First Floor, HSL Buildings
437 Warrington Road
Rainhill
Merseyside L35 4LL

Preface to First Edition

In the UK, home ownership has traditionally been the ambition of a very large section of the population. The concept of the property-owning democracy is something that has been fostered by a large number of British politicians for years. Although tax concessions on housing are now being slowly eroded, there is no doubting that home improvements will continue to provide work for a large number of tradesmen builders, home improvement companies and professionals.

The basic skeleton of this book was created many years ago when I prepared an "in-house" manual containing my standard specifications, standard details and standard letters for use in my office. Although produced long before the introduction of BS 5750, the reason for creating the manual was the same. I wanted to ensure that all drawings produced in my office were prepared to a uniform standard and that procedures could also be standardized. The first edition of this book was published by *Building Trades Journal* as part of their Practical Guide series and delighted under the snappy title of *A Practical Guide to Single Storey House Extensions*.

Although the basic principles described in the previous publication remain valid, major revisions to the Building Regulations have now made some of the technical information in the original very out of date. In addition, since the first edition was published, I have been pleasantly surprised to receive several telephone calls praising the original but asking if I ever intended to produce a more advanced version of the book. Thus motivated, I have taken the opportunity to expand the contents.

The book is still aimed at the same target audience, namely:

(a) Smaller building companies/tradesman builders.
(b) Draughtspersons/junior architectural technicians.
(c) Surveyors/building consultants.
(d) Students.
(e) Householders considering altering their homes, who want to "read around" the subject.
(f) The keen DIY enthusiast.

Whilst the tradesman builder might be an expert in their own speciality, in my experience, most do not know a great deal about planning and building control procedures. The book should therefore provide them with a useful guide, especially concerning matters such as Permitted Development and the effects of modern Building Regulations.

Students, technicians and consultants will hopefully find the technical details of assistance when preparing their own plans.

For the householder, this book will hopefully explain the basics of the planning and building control system in England and Wales and give them an insight into simple building construction. Then, if and when they engage a building surveyor to design their extension, they will understand what work is being carried out on their behalf. I have stressed basics, because the book is only intended to cover simple domestic situations.

For the DIYer, a challenge is always a challenge and I have no doubt that this book will provide enough information for the avid DIYer to prepare his own plans for the simpler type of home extension and submit them to their local authority for planning and building control approvals. Naturally enough, a little knowledge can be a dangerous thing, and I would not envisage that a DIYer would attempt to design a major building.

Andrew R. Williams FRICS, FCIOB, FBEng, FIAS, MIBC
Andrew R. Williams & Co.
Chartered Quantity Surveyors
Corporate Building Engineers
First Floor, HSL Buildings
437 Warrington Road
Rainhill, Merseyside, L35 4LL

Acknowledgements

The author wishes to thank David Tierney (Principal Building Control Manager of Halton Borough Council), Ian Davis (NHBC Director of Technical Standards), Steve Le Guen and Leonard Fernley for their help and assistance whilst preparing the manuscript for this book.

As many of the technical illustrations were derived from manufacturers' technical catalogues, the author also wishes to thank Ibstock Building Products Ltd., Owens Corning (formerly Pilkington Insulations), Willan Building Services Ltd., Marley Building Materials Ltd., Alfred McAlpine Minerals (Penrhyn Slate Quarries) and WL Computer Services. Last but not least, the author wishes to thank Halton Borough Council and St. Helens Borough Council for permitting some of their standard forms and documents to be reproduced as examples.

The photograph on the front cover shows a typical extension under construction. The photograph was supplied by Mr M. and Mrs A. Aldridge (of Cheshire) and Mr D. Glover (Building contractor), and has been reproduced with their permission.

ADDITIONAL ACKNOWLEDGEMENTS FOR SECOND EDITION

In addition to the acknowledgements given in the first edition, the author wishes once again to thank David Tierney (Principal Building Control Manager of Halton Borough Council) for his continuing assistance, Building Product Design (formerly William Building Products) Richard Smith for his help and assistance and Peter Barr of SIPBuild for his assistance with the section on SIPs.

PART ONE
Introduction

1 A general guide to drawing the plan

GENERALLY

Figure 1.1 indicates the basic information that will be needed on most plans. The plan is not meant as a solution to all situations but to provide good grounding. If you produce a drawing something like this you are well on the way to getting approval to your scheme. For a typical house extension plan, the following minimum details will probably be required by Planning and/or Building Control Departments (full details concerning Planning and Building Control are provided later):

(a) Plan of the existing (for a ground floor extension, a plan of the ground floor will probably suffice) – minimum scale 1:100 (1:50 is preferred) with drainage details.
(b) Plan of proposed (for a ground floor extension, showing existing house and extension) – minimum scale 1:100 (1:50 is preferred) with drainage details.
(c) Section through building – minimum scale 1:100 (1:50 is preferred).
(d) Existing rear, side and/or front elevation (as applicable) – minimum scale 1:100 (1:50 is preferred).
(e) Proposed rear, side and/or front elevation (as applicable) – minimum scale 1:100 (1:50 is preferred).
(f) Site location plan. (**Note**: Most local authorities now require an ordnance survey (OS) sheet.)
(g) Block plan (sometimes the site location plan will suffice).
(h) Full specification of materials to be used, cross-referenced to the drawing.

The plans should also show the following:

(a) The position of the ground levels.
(b) The position of the damp-proof courses (DPC) and any other barriers to moisture.
(c) The position, form and dimensions of the foundations, walls, windows, floors, roofs and chimneys.

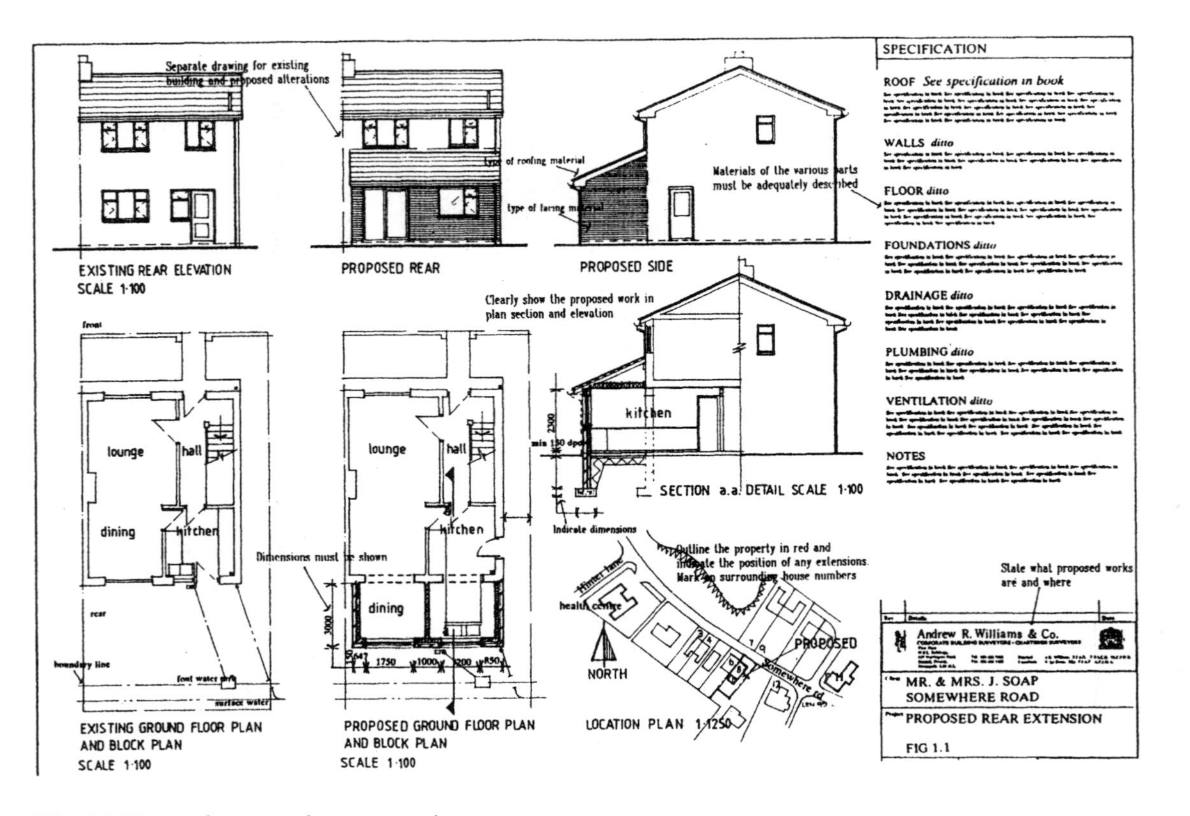

Fig. 1.1 Plans of proposed rear extension.

(d) The intended use of every room in the building(s).
(e) The provision made in the structure for protection against fire and for insulation against the transmission of heat and sound.
(f) The provision to be made for the drainage of the building or extension.
(g) The existing features of the site including any trees, outbuildings, those parts which will be demolished and be in sufficient detail to give a clear picture of any new building. Where existing and new works are shown on the same drawing, new work should be distinctively marked using hatching. The materials to be used in the external finish of walls and roofs and their colour should be indicated on the drawings (this is part of the specification on the drawing). On small works, the term 'to match' existing is normally sufficient when dealing with roofing and brickwork.

THE BLOCK PLAN/LOCATION PLAN

The location plan can be drawn on the main plan at a scale of 1:1250. As indicated above, alternatively an ordnance survey sheet can be submitted. The location plan should also show

(a) The size and position of the building(s) and its (their) relationship to adjoining buildings.
(b) The width and position of every street adjoining the premises.
(c) The boundaries of the premises and the size and position of every other building and of every garden, yard and other open space within such boundaries.
(d) When submitting an application for planning permission the boundaries of the application site should be edged in red.

SCALES GENERALLY

As indicated above, elevations and floor plans are normally to a scale of not less than 1:100. It is recommended that scales of 1:50 be used wherever possible as this makes the plans clearer, particularly when drawing sections.

LARGER PROJECTS

Obviously, the more ambitious the proposals, the larger number of plans, elevations and sections that will be required on each full drawing. If the extension or alterations affect two floors, then plans for both floors must be provided (proposed and existing). In theory, the plans should show every floor and roof of the building and a section of every storey of the building. In practice, for smaller projects, it is normally acceptable merely to show details of the floors being altered.

CHECKLIST OF ITEMS OFTEN MISSED OFF PLANS

In Appendix A, I have included a short checklist of items that are often missed off plans or sometimes not addressed at all by the draughtsperson/ CAD operative when preparing plans. The list is not intended to cover every eventuality and could doubtless be extended.

STANDARD SPECIFICATION

In Appendix B, I have included a typical specification. It is very comprehensive and some might say it has an element of 'over-kill' about it. Because some items may not be applicable, it is usual to delete the inappropriate items when preparing a specific plan. As it would be extremely tedious to have to handprint this document onto every drawing, the simplest way of reproducing the standard specification on the drawings is by

(a) Turning it into a booklet form and cross-referencing the main plan.
(b) Creating a master specification plan.
   (i) You can make the master specification a full drawing in its own right, and copy it and send it in with the project drawing.
   (ii) You can include the specification on the main drawing.

THINKING IT THROUGH

Generally

Although it may seem fairly obvious, whether you are the builder/ homeowner or the surveyor who has been called in to prepare plans, it is essential from the very outset to decide exactly what is wanted. Before visiting a client, I often suggest that they make rough sketches of what they want to do to their property so that it is possible to swiftly discuss the practicality of a proposal once I arrive. It has to be borne in mind that even though most homes lend themselves to extension, you need to weigh up many factors before finalizing the design. I have listed a few for consideration.

(1) Does the improvement justify the expenditure?
   It is possible to enlarge a property so much that the money invested is unlikely to ever be recovered if and when it is finally sold (e.g. someone enlarging a three-bedroom house into a six-bedroom house in an area that is principally composed of smaller properties is likely to be in this position). I have actually been faced with situations where I have advised potential clients to move rather than extend, for exactly the reasons given above. Admittedly, I have had to forgo the odd

commission, but I believe that honesty pays. I certainly would not wish to slip to the level of a high-pressure salesperson and advise people to carry out a project that I believed to be foolhardy.

(2) As well as the pluses are there minuses to the proposals?
Take as an example the family that want a dining room built onto their house. An extension of this nature *that* could only be reached via the existing kitchen is not as desirable as a dining room off a hallway. Kitchens can be untidy and smelly places whilst cooking is going on. Having to take guests through a working kitchen in order to reach the dining room might, in the long term, become an embarrassment. What if you wanted to impress your new boss or a highly critical relative? Having to drag them past pots and pans is far from ideal.

(3) What will the effect be on the external appearance of the house?
If you drive around your home town for a short while, you will come across many examples of badly thought-out schemes. How about these for starters:

(a) The two-storey side extension to a normal two-storey house. The original house has a tiled roof but the extension has a flat roof. Admittedly, most Planning Departments nowadays reject applications like these, but some have been built. Very few people would describe this type of extension as beautiful, but it is amazing how many were built this way just to save money. (My office has recently designed a series of new tiled roofs to replace leaking flat roofs in our area when the owners suddenly realized that the original flat roof had been a false economy.) No doubt one or two planners let out a cheer as they were replaced.
(b) The loft conversion that is so dominant that it completely destroys the existing roof line.
(c) The extension built of badly matched bricks. (It saved money but the owner now lives to regret the penny pinching.)

(4) What will be the effect on adjoining properties?
I have known several situations where good neighbours fell out merely because one ignored the wishes of another.

(5) Is the proposed extension oversized and badly sited?
Badly designed extensions can result in loss of sunlight/outlook both for the instigator of the extension and the adjoining owners.

As people do still ignore obvious design considerations, some sort of control is needed, and that is the subject of the next chapter.

2 The need for control

WHO CARES WHAT THE NEIGHBOURS THINK?

Do you remember the newspaper articles and TV reports about the man who decided that a house was not really a home unless it had a very large model fish on the roof and a secondhand armoured fighting vehicle in the front garden? I gather from the media reports that his neighbours are not exactly pleased with his 'follies'.

COMMON SENSE IS NOT ALL THAT COMMON

There is an old saying that common sense is not all that common. Do you remember the newspaper story about the man who decided that it might be nice to have a basement under his house? You know the one, he laboured away all summer and as he was digging, he came across some obstructions in the ground (in the trade, we call them foundations), so he dug them up and the house, naturally enough, collapsed.

Knowing newspapers it might have been an apocryphal story, but I have seen members of the public doing some extremely stupid things over the years, either because they did not know any better or because they had decided that the rules did not apply to them. Then there is the person who is doing something as a 'matter of principle'. Solicitors love them because as they say 'principles cost money'.

Unfortunately, there will always be some members of the 'it's-my-property-I'll-do-what-I-want-with-it' brigade in every area, and that is why the various rules and regulations become more strict as the years go by.

At this point, I am going to toss in a personal anecdote as a further illustration. Many years ago I was asked to prepare a set of plans for an extension on a property. As the enquiry was generated by a tradesman builder that I knew reasonably well, we set off together to visit the house in question. Once under way, Tony, the builder, explained that he had given the owner a budget quotation for the work based on a cost per square foot of floor area, and wanted to come with me to ensure that he had a good look around before firming up his price.

Once we arrived and had been invited in, Tony gave me a puzzled look and then whispered 'Funny house this!', and then went back to staring at the patch of new plaster on the ceiling. While Tony had been eyeing up the plaster, I had been weighing up a timber post in the middle of the living room. Then I realized that there was a sink on the far side of the living room. It was at that point that Tony drew my attention to the fresh plaster patch. This had, without any doubt, been the result of repair works following the removal of the chimney breast in the living room.

Being concerned about the structural integrity of the property, I said, 'Have you put in RSJs (rolled steel joists) to support the remaining chimney breast upstairs?' The owner looked at me and shook his head. It was at that point that I glanced around the rest of the house with more enlightenment. Only moments later, the truth dawned! This silly man had demolished all the load-bearing partition walls downstairs; the entire weight of the first floor was supported on the 4 in × 4 in (100 mm × 100 mm) wooden pole and the external walls. It was then that I warned the owner that I considered the whole property structurally unsound. I think that I upset him because the gentleman in question made it clear that if it had not been for his wife's insistence he would have built the new extension himself as well.

As far as he was concerned, he could see no reason for submitting plans to the council. It was all just 'red tape' designed to keep overpaid civil servants in the lap of luxury. He could not see anything wrong in his previous 'do-it-yourself' attempts. It was at that point that Tony and myself both came to the same conclusion: we did not want anything to do with this character.

However, what if you were the next door neighbour? Bearing in mind that this house was a mid-terraced property, would you like to live next door? When the house collapses, as it will in the fullness of time, and damages other houses, what will happen? Maybe someone might even be killed!

It is people like the man above, who do not know what they are doing, or who skimp to save money, that make controls so essential. Such examples also prove the need for competent designers and builders. People who have no knowledge of building construction and are not prepared to learn should not dabble.

3 Planning and building control

INTRODUCTION

As I have indicated in the previous chapter, some control is needed over 'development' and most sensible and informed people appreciate this, even though at times they become frustrated by the fact that the procedures seem to take so long.

It should also be obvious that because of the actions of a few unthinking people in the past, there is now very little that can be done to a property without having to consult the local authority (the local council). Thinking people would probably also agree that if the regulations were relaxed that the 'fringe element' would return to their old ways. To some, having to conform to rules is seen as an infringement of their civil liberties, but as I have tried to indicate in the previous chapter, control over 'development' is no bad thing.

Although there are exceptions (which will be discussed later in more detail), no work should commence unless the necessary approvals have been obtained from the

(a) Building Control Department (or an Approved Inspector)
(b) Planning Department/Planning Authority. (**Note**: As explained in later chapters, some alterations and extensions are Permitted Development and do not require planning approval.)

In some areas, such as conservation areas, additional approvals are also required. (Practically speaking, ignoring the time limits that both departments are supposed to work to, in my 'patch', it normally takes at least two months to go through all the procedures, but it can take longer.)

In the main, the Planning and Building Control Departments come under the control of the local authority. However, sometimes the planning function will be either partially or totally under the control of a non-elected body such as a New Town Development Agency. With privatization now being the new buzzword, councils have also lost some of their authority over Building Control now that Approved Inspectors have arrived on the scene. (I will describe the role of the Approved Inspector in a bit more detail later.)

Generally speaking, it would be unwise to commence work until plans have been drawn up, copies have been forwarded to the Building Control Department and Planning Department using the prescribed forms and both departments have given their written approval to the proposals.

WHAT IS THE DIFFERENCE BETWEEN PLANNING AND BUILDING CONTROL?

The two terms are of course a form of jargon. Like all professionals, local authority personnel have developed their own shorthand and they seem to surround themselves in their own form of 'newspeak'. This can cause confusion for lay people. Quite often, people telephone my office and say that someone from the Planning Department has told them something but once asked for more detail, it becomes obvious that it was probably a Building Control Officer (BCO) who provided the information and vice versa. So, what is the difference? The easily understood answer is the Planning Department is interested in what the building looks like and Building Control wants to know if it is structurally sound.

To put it another way, the Planning Department will concern itself with such matters as the number of off-street car parking spaces that will be left once a house has been extended. It will be highly concerned if it becomes apparent that a proposed house or extension will occupy an excessively large amount of land available on the plot (the overdeveloped site), or if the proposals are likely to overcrowd a neighbouring property. Likewise, if a proposal is likely to 'overlook' a neighbour, the department will be concerned.

Beware! The criteria that Planning Departments apply to a problem might differ considerably from area to area. Most local authorities have their own guidelines and most will provide copies of their policies if asked to do so. As a glaring example of differing local policy decisions, in my 'patch' there is one local authority that only allows 3 m deep rear extensions to most semidetached and terraced properties. The adjacent authority allows 4 m deep rear extensions. As the boundary between the two authorities is often the centre line of a road, one should consider the difficulty that a consultant surveyor might have explaining to a prospective client who lives on the 'wrong side of the road' that he or she will not be allowed to do what a neighbour has done close by.

Building Control will concern itself with the structural stability of the proposal (i.e. are the walls strong enough to support the loads being placed on them), if the building is watertight and whether it will cause a danger to the present or future occupants or surrounding properties if there was ever a fire in the building.

PART TWO
Householder Developments

4 House extension policies and The Party Wall etc Act 1996

PREAMBLE

I have described house extension policies at the end of this chapter. I have also described Permitted Development rules in Chapter 5. However, even as I write I am conscious that things are likely to change.

And, if the *Daily Telegraph* article of 23 May 2007 is right, the planning reforms 'will extend chaos'. The same article in the *Daily Telegraph* also asks a good question. 'Does the government really believe it can make the planning system less onerous by inviting us to make agreements with our neighbours?' The journalist then goes on to point out that British families have a reputation for disagreeing with whatever a neighbour does with their house.

So will the new rules be easier to understand or will chaos ensue? Well – we shall see, won't we?

The New Planning Proposals/Revisions to Permitted Development

As indicated above, at the time of writing this section of the book, Communities Secretary Ruth Kelly has just announced that there are proposals afoot to do away with planning permission for 'minor home extensions'.

Although the full details have yet to be revealed, one can't help but think that some of the talk is once again just government 'smoke and mirrors' because if you read on you will note that under the present system, a large number of home extensions do not require planning permission because they are deemed to be Permitted Development (PD).

BEWARE – Just because a proposal is PD for Planning purposes, does not exclude it from Building Control procedures.

On the other hand, maybe a change is overdue because unfortunately certain authorities seem reluctant to concede that *any* minor works are PD and find excuses for demanding a planning application.

Ruth Kelly's proposals also included 'impact testing'. In other words, if someone proposes to carry out alterations/extension they will be required to assess the likely effect on the neighbour's property. However, once again, this is not really as new as it looks because as I have indicated later in this chapter, most local authorities publish domestic extension guidelines which indicate what is acceptable and what is not acceptable to the local Planning Department.

Obviously, at this stage, it is difficult to guess at what will finally be approved by Parliament. However, there are outlines already published that indicate the reasoning behind the proposed changes. They are:

(1) The Steering Group Report has decided that the existing General Permitted Development Order is too complicated for normal people to understand.
(2) The Steering Group Report wants the new Permitted Development Order to have explanatory guidance in plain English.

The Steering Group also wish to ensure that the new Permitted Development Order for household developments will ensure:

(3) There is clarity, simplicity and consistency.
(4) The need for planning permission is proportionate to the impact of the development.
(5) The number of (household) planning applications is kept to minimum.
(6) That the regulations are and remain relevant to new techniques and changing lifestyles.
(7) The new order would set out common categories of household development (e.g. roof extensions, conservatories, side extensions). The current 'volume' rules (discussed in later chapters) would be scrapped.
(8) That householders are encouraged to consult their neighbours before submitting a proposal to reduce the number of neighbour objections to an application.

So when is this all likely to happen?

This is a question that cannot be answered. When I have spoken to local Planning Officers about the proposals the answer is usually the same. . . . It's a long way off. This is why, later in this chapter I have explained current house extension policies and have provided information on the way that the system works at the moment.

THE EFFECTS OF THE PARTY WALL ETC ACT 1996 ON ALTERATIONS AND EXTENSIONS

The Party Wall etc Act

The Party Wall etc Act (the Act) came into force on 1 July 1997 and applies throughout England and Wales.

Prior to the Act being passed, the only part of England and Wales that had party wall rules was the London area. Ignoring the effects of Building Control/Conservation Areas etc., prior to the Act, anyone applying for planning approval in England and Wales (excluding London) sent in their plans and once approved could start work.

Since the Act, life is not so simple.

However, this should not be the cause of doom and gloom because the Act provides a building owner with additional rights that he never enjoyed before as long as all the relevant notifications are issued to the neighbours in good time.

How does the Act affect surveyors/architects or draughtspeople

When dealing with small domestic works (and the surveyors/architects or draughtsperson is not involved with site supervision), the Act has, in my opinion, had very little effect on design professionals because the Act requires the *owner* of the property being altered/extended to issue the relevant notices on their neighbours.

So if we can still draw our plans up exactly the same way that we did before the Act came into force, why am I mentioning the Act?

The answer is simple, although we can still draw our plans up exactly the same way that we did before the Act came into force, most clients would expect their professional adviser to warn them of the effects of the Act. Judging by the conversations that I have had with members of the public since the Act was passed (and one or two letters from solicitors), some people obviously think the Act is a way of preventing your neighbour from carrying out works to his/her home.

The truth is totally different.

The Party Wall etc Act is an enabling Act which is designed to prevent or resolve disputes between neighbours. If all else fails, the Act provides a means of resolving disputes (i.e. The appointment of a party wall surveyor).

Unlike most of the Acts of Parliament, this Act does not provide juicy pickings for the legal profession. If a dispute arises and cannot be resolved, a party wall surveyor is appointed (or party wall surveyors are appointed) to negotiate a solution, not a solicitor. Admittedly, there is nothing to prevent a solicitor setting himself/herself up as a party wall surveyor but I would suggest that most solicitors would be lacking in the relevant construction training to adequately carry out the function properly.

Also note the *etc* in the name of the Act; The Act doesn't *just* cover work to party walls.

WHAT TYPES OF WORK ARE COVERED BY THE ACT?

(1) Various works that are going to be carried out directly to an existing party wall (the common wall between a semi-detached house or a terrace) or structure.
(2) New building at or astride the boundary line between properties.
(3) Excavation within 3 or 6 m of a neighbouring building(s) or structure(s), depending on the depth of the holes or proposed foundations.

PARTY WALLS

The Act recognizes two main types of party wall.

Party wall type (a)

A wall is a 'party wall' if it stands astride the boundary of land belonging to two (or more) different owners.

A wall is a 'party fence wall' if it is not part of a building, and stands astride the boundary line between lands of different owners and is used to separate those lands (e.g. a garden wall). *This does not include such things as wooden fences.*

Party wall type (b)

A wall is also a 'party wall' if it stands wholly on one owner's land, but is used by two (or more) owners to separate their buildings (e.g. where one person has built the wall and then another person has butted their building up against it without constructing their own wall).

Party structure

The expression 'party structure' is a wider term (e.g. a wall or floor partition separating buildings or parts of buildings approached by separate staircases or entrances – Flats).

EXPLANATORY BOOKLET

Obviously, my description above is only a very brief discourse on the Act and I would advise my readers to obtain a copy of *The Party Wall etc Act*

1996: Explanatory Booklet (a free government publication), which explains the duties of householders who wish to alter/extend their property.

THE EFFECT OF THE ACT IN GENERAL TERMS

As indicated above, in general terms the Act relates to the following:

(1) New walls (of an extension or a fence wall) that touch the boundary (line of junction);
(2) Work to party walls or structures adjoining party walls;
(3) Excavations or foundation works which are within 3 m (or in some cases 6 m) of neighbouring structures, where it is the intention to excavate lower than the underside of the foundations of the neighbouring property.

NEW WALLS ON THE LINE OF JUNCTION

If you refer to Fig. 8.11 (the uncooperative next door neighbour) you will note that I suggest leaving a 150 mm (6 in) gap between the extension and the boundary (line of junction) if there is an awkward neighbour.

However, strictly speaking, this ploy is no longer necessary because the Act indicates that makes it totally legal to construct projecting foundations under a neighbours land.

Where an owner is proposing to build on the boundary (line of junction), the neighbour has to be formally notified as per the sample letter below:

TYPICAL EXAMPLES OF THE PARTY WALL ACT IN PRACTICE

Let's say that Mr Extender wants to build an extension to his property.

The extension in question is built up against the boundary line. Applying the rules of the Act, Mr Extender must send a letter to his neighbour, Mr Adjoiner, a letter which would read something like this:

Mr Adjoiner,
12, Somewhere Road,
Tipham.

Date.........

Dear Mr Adjoiner,

The Party Wall etc Act 1996 – Line of Junction Notice

A the owner of 10, Somewhere Road which is adjacent to your premises at 12, Somewhere road, I notify you that in accordance with our rights under section 1 of the Party Wall etc Act 1996, I intend to build at the line of junction between our two properties.

We enclose Plan ARW1 which shows the wall of a proposed rear extension built wholly on my land. Under the right given by section 1(6) of the Party Wall etc Act, it is intended to put projecting foundations under your land. I intend to start work on one months notice.

In the event of a dispute between us under the Act, would you be willing to agree to the appointment of Andrew R. Williams as the agreed party wall surveyor.

If the answer is no, please let me know whom you would like me to appoint as your surveyor.

Yours sincerely,

Mr Extender

WORK ON THE PARTY WALL

If you carry out work to a party wall for instance,

(1) Cut into a party wall to take the bearing of a beam, or,
(2) Rebuild or raise the height of a party wall, or,
(3) Underpin a party wall.

Then you have to inform your neighbour of your intentions at least TWO months before the work starts.

THE COST OF PARTY WALL SURVEYORS

Coming back to Fig. 8.11 (the uncooperative next door neighbour), the main reason for still adopting the 'set back' of 150 mm is that if the neighbour refuses to co-operate and a party wall surveyor is appointed, the costs fall on the person wanting to build an extension/alter their property. The main problem with party wall disputes is that they cost money for the person wanting to alter their property. Once a firm of party wall surveyors that I know of charged around £1400 to settle a simple domestic party wall issue.

LOCAL AUTHORITY HOUSE EXTENSION POLICIES

As indicated above, most councils issue guidelines for extensions to domestic properties in their area. For simplicity, with Halton Council's permission, I have reproduced a typical householder's design policy (see Appendix D).

This has only been reproduced as a guide. When designing any building, it is essential to obtain the policy document issued by the local authority having control over the area in question. The policies differ from council to council. As the information is fairly self-explanatory, I would suggest that you study the document as a typical example of the sort of requirements that are normal.

Note: In Chapter 5, I have provided information concerning Permitted Development. If the extension constitutes 'Permitted Development' then the council (from the planning point of view) has virtually no control over the design. In other words, generally speaking, design policies can only be enforced if planning approval (permission) is needed.

My advice is to check to see if you need planning permission first.

It is often possible to get a situation arising where the local Planning Department would refuse a full planning application because their design criteria had not been implemented whilst simultaneously agreeing to the alteration under Permitted Development rules.

GENERAL NOTES ON HOUSEHOLDER EXTENSION POLICIES

The following comments are made based on what is normal in my area; in my experience, the policies of most Planning Departments cover the following:

Harmonizing – most councils wish to ensure that all materials used in an extension match as closely as possible those of the existing structure. (They do not want it 'to stick out like a sore thumb'.)

Overlooking neighbours – in order that the 'nosey neighbour' syndrome can be avoided, most Planning Departments have created guidelines and set down minimum distances between parallel windows at the front, side and back so that 'overlooking' is minimized (study the Halton Policy as an example).

Dominance – in my area, most councils require an extension to be 'an unobtrusive and subordinate' addition. As a further safeguard to privacy most councils require that when an extension is built, there should be a minimum distance between the rear of the extension and rear boundary of the property.

Remaining garden space (overdevelopment) – some councils have a requirement that, once an extension is completed, a minimum area of free garden space is left at the rear.

Single-/two-storey extensions to the rear – most Planning Departments have requirements concerning how far an extension can project from the rear of a property when the proposal is fairly close to one or more side boundaries with a neighbour. Some have a set distance for both upper and lower storeys, others create formulas. The Halton Policy uses a 45 degree rule.

Side extensions – most Planning Departments are concerned that side extensions should not create a 'terracing effect' (i.e. if two neighbours who own semidetached houses both extend them at ground and first floor level, then the block ends up looking like a terrace and has all the problems associated with terraced properties). For example, lack of access to the rear, problems with sound transmission, people living 'cheek by jowl'.

Existing building lines – most Planning Departments try to restrict front extensions (other than very minor ones such as porches). Thus someone who seeks to obtain permission for a very dominant front extension is likely to receive a refusal.

Flat roofs – flat roofs at second floor level are not normally looked upon with favour. In one authority in my area, flat roofs to single storey extensions are actively discouraged.

Dormer extensions – where a loft is converted, and a dormer window is proposed, most Planning Departments will refuse an application if the new construction projects above the ridge line of the existing property.

Garages and parking generally – to restrict on-street parking, most councils now require that adequate car parking is available once the proposed extension is completed. A proposal that completely uses up all available car parking space (or a substantial part of it) will normally be rejected. Most councils do not like garages to project in front to the foremost part of the house unless built with a porch.

There is also usually a requirement that a driveway in front of the garage should be large enough to accommodate a car. In my area, most councils require 5.5 m to be provided in front of the garage door. This requirement is to ensure that the owner can fully pull off the road before opening the garage door. It makes good sense!

Unsightly bonding on front elevations (see Fig. 8.7) – in order to avoid cutting and bonding to brickwork being visible on front elevation most planners require a 'setback' of at least 112 mm where an extension is attached to the dwelling. (**Note**: If the property is subject to 'anti-terracing' rules, the setback might well be a metre.)

5 Is planning approval necessary?

PERMITTED DEVELOPMENT

As I have stressed the need for control in the first few chapters, the question 'is planning approval necessary?' may seem strange (as opposed to building control approval which is another matter). The simple fact is though, not all alterations require planning approval as long as certain criteria are met.

Although the Town and Country Planning Acts give local authorities the power to control 'Development', their powers are restricted in certain circumstances. If the proposals happen to fall within categories called Permitted Development (PD), then the Planning Department have to accept that the works can be carried out, whether they like it or not (remember, building control approval may still be required). In England and Wales, what constitutes PD is defined in the Town and Country Planning General Development Order.

It may seem strange that there should be what is in effect a two-tier planning system, but it must be remembered that the restriction on the powers of the Planning Department is a common sense decision. If local council powers were total then the council offices would be flooded with all sorts of trivial applications which would quickly clog the system. No one would be able to do anything without needing to consult the local council. The PD system was designed to benefit both the local authorities and the general public.

For the designer and the homeowner, the point at which a home extension ceases to be PD and requires a full planning application is very important.

NOTES ON PERMITTED DEVELOPMENT

Generally

As explained above, what constitutes PD is defined in a General Development Order (GDO) issued by the Secretary of State for the Environment and covers a wide range of 'Development'. The comments below apply only to domestic property and, like all things, GDOs are revised from time to time. (Beware – Article 4 of the GDO provides local authorities with the power to give a direction that Permitted Development rights can be suspended in certain circumstances. It is prudent to make enquiries with the local authority when proposing to deal with what is obviously a fine building to see if any restrictions are in force. This caution should also be exercised when dealing with works in national parks, areas of outstanding beauty, conservation areas, buildings that are likely to be listed or where experience indicates that a previous planning consent has removed PD rights.)

Permitted Development is divided into many classes. (The 'rules' listed below are my rather simplistic interpretation of the GDO. For more information, you must obtain a copy of the current GDO and read it fully.) For convenience, I have divided PD as applicable to domestic alterations into the following three types.

(1) porches
(2) dormer windows to lofts
(3) other extensions.

1. Porches

Porches can be built without planning approval (**Note**: Check with local authority if the house is in a national park, an area of outstanding beauty, or a conservation area, or if it is listed, or if a previous planning approval may have removed PD rights.) as long as

(a) Ground floor area (measured externally) does not exceed $3\,m^2$.
(b) No part of the porch exceeds 3 m in height.
(c) No part of the house (or the porch when constructed) is closer than 2 m from a highway.

2. Dormer windows to lofts

Dormer windows to loft spaces can be constructed without planning approval (but check with local authority as for porches) as long as

(a) The dormer would not project above the highest part of the roof (e.g. ridge line).
(b) The dormer does not face a road.
(c) Where a loft conversion would not add more than

  (i) $40\,m^3$ to a terraced house
  (ii) $50\,m^3$ to any other house.

3. Other extensions

Other extensions can be built without planning approval (but check with local authority as for porches) based on the following restrictions.

A. *Terraced houses and houses on Article 1(5) Land*

Subject to the other rules regarding limitations on PD given below, terraced houses and houses on Article 1(5) Land, i.e. land within

(a) a national park
(b) area of outstanding natural beauty
(c) conservation area created under section 27 of the Town and Country Planning Act
(d) area of natural beauty.

can be extended by $50\,m^3$ or 10 per cent of their floor area (whichever is greater) without planning permission subject to an upper limit of $115\,m^3$. (**Note**: Where a proposed extension has a pitched (tiled or slated) roof, the cubic capacity of the roof has to be taken into account as well.)

B. *Other types of houses*

Subject to the rules regarding limitations on PD given below, detached and semidetached houses can be extended by $70\,m^3$ or 15 per cent of their floor area (whichever is greater) without planning permission subject to an upper limit of $115\,m^3$. (**Note**: Where a proposed extension has a pitched (tiled or slated) roof, the cubic capacity of the roof has to be taken into account as well.)

Limitations on Permitted Development for type 3 ('Other extensions')

The following limitations are applied to PD rights when dealing with 'type 3' extensions. (Speaking in very general terms, in practice, PD rights can only be effectively used for ground floor extensions which would not affect the existing main roof, but there are obviously always exceptions, as would be the case if you were dealing with a dwelling that was several storeys high.)

(a) No extension must be built higher than the existing house.
(b) The PD rules cannot be used in front garden situations, unless the front garden does not face a road or highway or the length of garden left after building the extension would exceed 20 m (approx. 66 ft). (**Note**: The term 'highway' does not just cover roads, it also covers public footpaths and bridleways. It presumably could also be interpreted as

including an accessway sandwiched between rows of 'back to back' terraced houses.)

(c) The PD rules cannot be used where an extension faces a road or highway on any side of the house, unless the length of garden left after building the extension would exceed 20 m (approx. 66 ft). (**Note**: Take this rule into account when dealing with houses that have roads or highways front and back and/or side. As far as PD is concerned if a house has roads/ highways on more than one side, then unless the length of the garden is big enough (20 m) planning permission must be obtained.)

(d) The PD rules cannot be used where an extension is within 2 m of a boundary and any part of the extension within that 2 m is over 4 m in height.

(e) The PD rules cannot be used where more than 50 per cent of the existing garden area would be used up by the extension. (**Note**: When dealing with terraced houses with no front garden and only a small yard at the back this is a very real restriction.)

(f) The PD rules cannot be used for 'type 3' works where there would be an alteration to any part of the existing roof. (My interpretation of this rule would be that PD might not be allowable if the proposal concerned the alteration of a bungalow and it is necessary to intersect with the existing pitched roof.)

Five metre rule

When calculating the cubic capacities for 'other types of houses' (3B above), the erection of a detached garage in the curtilage of the site is only treated as being an enlargement to the house if it is within 5 m of the house (or the proposed new extension to the house).

If it is possible to disregard the garage and its cubic capacity, then the owner of the property 'gets another bite of the cherry'.

FURTHER NOTES ON PERMITTED DEVELOPMENT

Obviously some people might, if following the above rules without much thought, decide to extend their house by the Permitted Development limit one year and then try to do the same thing the year after and so on. *You cannot do this*. Once you have used up your permitted limit – that's it!

However, after saying that, there are always exceptions. The date 1 July 1948 is an important date to remember when dealing with PD. If your (or your client's) house has obviously been extended and it can be reasonably substantiated (say by examining the deeds of the property in question) that the extensions were carried out prior to 1 July 1948 (and no other extensions have been built after that date to use up the PD limit), then the

PD allowances discussed above can be used to the full. In this case, the old extension is treated as being part of the original property.

One other important factor to take into account when using the benefits of PD, once you have a builder working on site, is that the builder is made aware of the rules that they must comply with. Some builders, if left to their own devices will build to suit themselves and if they exceed the stipulated limits and the Planning Department discover the discrepancies, they will be within their rights to demand a retrospective planning application. If for some reason they do not accept the alteration as being acceptable, the extension may need to be demolished or modified.

USING YOUR PD RIGHTS SENSIBLY

Generally

As indicated in Chapter 4, Planning Departments have design policies for their areas. In this chapter, I have indicated that these policies do not have to be applied when a proposal is deemed 'Permitted Development'.

However, I would not wish it to be thought that I am actively encouraging the populace as a whole to flout planning policies because in the main they have been created to protect the public (Chapter 2).

Purely as an example of how the system can be used sensibly, I intend to use another anecdote. A client that I once did work for wanted a 6 m long extension to his house. Under the policies applicable in the area in question, he would only have been allowed to build a 4 m long extension. When I explained that a 6 m extension would contravene the council guidelines, this upset him because his neighbour had no objections to the extension and as proof he spoke to the man next door in my presence. He became even more upset when I explained that even though the neighbour might agree to the extra long extension, it was unlikely that the Planning Department would agree to it as it would create a precedent. When both parties told me they thought that that was a rather silly situation, I then explained that there might be a way around the impasse using their PD rights.

Playing the PD game

If you refer to Fig. 5.1(a), you will see that I have made a sketch of my client's estate as originally constructed. All the houses are virtually identical. In Fig. 5.1(b), No. 3 Somewhere Road has built a 4 m long extension as per council policy, Nos 1 and 7 have done nothing whilst at No. 5, I proposed that part of the extension will be 6 m long and part 4 m long. Figure 5.1(c) explains how I achieved this. We constructed the extension in two phases. The 6 m extension was just under 70 m^3 and was built first. Once this

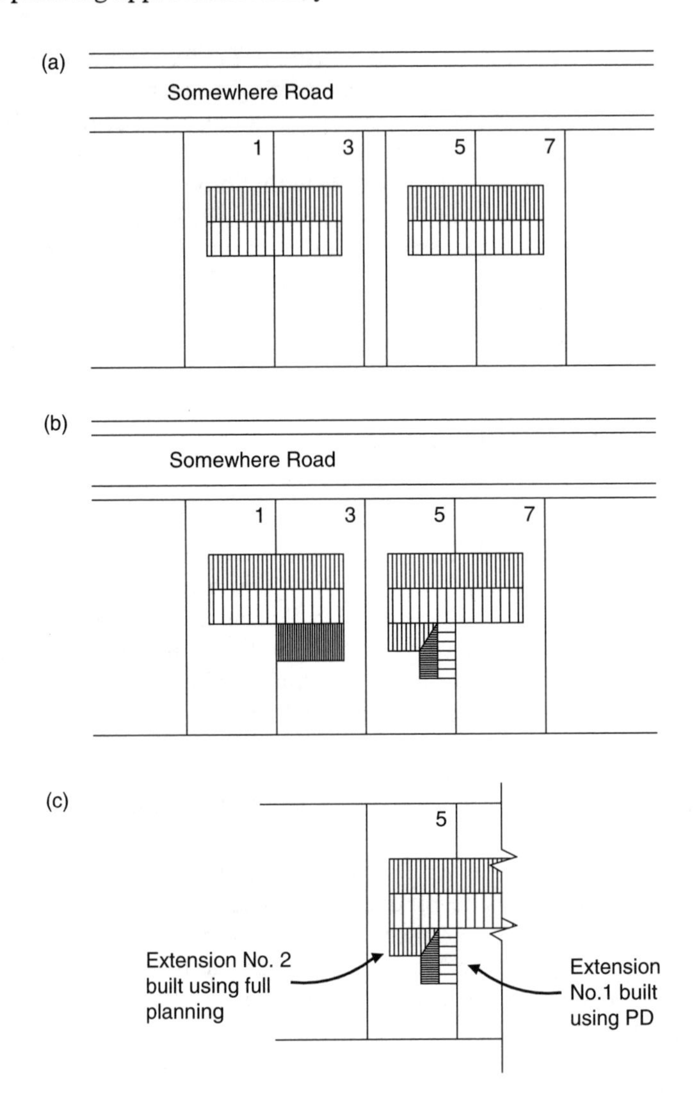

Fig. 5.1 Playing the PD game: (a) estate as originally built; (b) estate now; (c) how No. 5 used PD to his advantage.

had been built, we then applied for planning permission to build the 4 m section. It was quite legal!

So, as you can see, using your PD rights sensibly can help you in the following ways:

(a) As long as the property owner complies with the rules that govern PD, then he or she does not need to comply with local planning policy (e.g. if council policy only allows a 3 m rear extension but you want a larger extension, you can legally construct one as long as you obey say the 50 m^3/70 m^3 rules and all the riders).

Naturally, this is an anomaly that some Planning Officers dislike because once a proposal is permitted they are powerless to control the works. (**Note**: You still have to comply with the requirements stipulated by Building Control.)

(b) If the client does not have to formally apply for planning permission then normally the extension can be built very quickly because the planning process usually takes at least two months to complete and a confirmation that PD is accepted only normally takes two to three weeks.

(c) Under recent legislation, local authorities charge for planning applications. If PD is applied, the fees (fee scale can be obtained from your local Planning Department) will not have to be paid.

However, **beware**. I never take the risk of informing a client that the extension is automatically exempt from the Planning Regulations. When I am dealing with an application which involves what I believe to be PD, I usually write to the Planning Department along these lines:

Bigtown Planning Department,
Municipal Offices,
Somewhere Road,
Hyville

Date

For the attention of Mr Slowbend

Dear Sirs,

Proposed extension at 22 Bridge Street, Somewhere

Please find enclosed Plan No. . . . . . . This indicates a tiled roofed extension, not exceeding 70 m^3 with a height not exceeding 4 m.

There have been no other extensions to the property (as far as we are aware).

We are of the opinion that this extension constitutes Permitted Development and would request your confirmation/comments in writing.

We would confirm that we have applied for building control approval under separate cover.

Yours faithfully,

A. R. Williams
FRICS FCIOB FBEng

By doing this you cover both yourself and your client.

Remember, if you or your client proceeds without obtaining written permission, there is always a chance, slim though it might be, that there is a reason why PD is not allowed. Obtaining a letter from the local authority confirming PD status proves that you did consult with the authorities before proceeding and ensures that there can be no 'comebacks' at a later date.

Let the local authority tell you that the scheme is acceptable under the PD rules.

6 Applying for planning permission

ELECTRONIC PLANNING (e PLANNING)

Because the Internet is now easily accessible to most people, the government is encouraging planning authorities to provide an online service for applications and payment using the Planning Portal. Obviously, the hope is that this will speed up the planning process. Instead of having to rely on 'snale mail', the whole process is electronic. Instead of printing out large number of drawings, the plans will be sent to the local authority over the Internet as pdf images.

Obviously, submitting electronically also means that it allows the public easier access to the online Statutory Planning Register.

If you want to view this site first hand, type PLANNING PORTAL into the 'Google Bar' on your computer and the site will come up and you will be able to study the facilities available.

However, although electronic submission will very soon become standard practice, for the purposes of this book, I have used standard paper application forms in order to indicate what is involved.

Generally

You will realize from the last chapter that there are two alternative ways of dealing with the planning aspects of a proposed alteration/extension but I am going to go through the basics again.

Let us assume for the time being that you have prepared your plans and that you wish to submit them for approval. For a simple application there will be one of two alternatives, which are as follows:

(1) A letter to the Planning Department (enclosing a copy plan), requesting confirmation that the proposals are Permitted Development.
(2) A full planning application.

As I have covered the first alternative, let's assume that you need formal permission (approval).

THE FORMAL PLANNING APPLICATION

This chapter presumes that we are dealing with a fairly standard application and such things as tree preservation orders and the like do not apply. (This will be covered later.)

In order to obtain planning approval, your plans have to be submitted to the relevant local authority in your area together with a suitable application form; One supplied by the relevant local authority. (If you are using the Planning Portal described above the form comes up on the computer.)

It is worth bearing in mind that once your application has been submitted to the local authority, anyone can ask to inspect the documents lodged. If you submit a scruffy set of documents, it will do yourself no credit.

As I have indicated above, most local authorities have their own specially printed planning application forms. Each authority's forms vary slightly but the basic layout is the same. With the permission of Halton Borough Council, I have reproduced a typical Householder Planning Application Form (Fig. 6.1 (a) and (b)) which is fairly straight forward to fill in. (**Note**: Some authorities do not have Householder application forms. If this is the case with your authority, it will be necessary to fill in the full planning application form which is a little more complex.)

Parts 1 and **2 – These sections ask for the names and addresses of the applicant and agent**. The applicant is the person who wishes to have the works carried out (e.g. the homeowner). If you are working for someone (you have prepared the plans for someone other than yourself) and are looking after the submission, then you are the agent and you must insert your name and address in order that all correspondence will be directed to you. If you are not acting for someone (e.g. the application is for an extension on your own house), then there is no agent, so leave the box blank or cross it through. If you have a telephone number, provide it because if there are any queries, then someone can contact you easily and it helps to prevent delay.

Part 3 – This section asks for the address applicable to the application. This address is not necessarily the same as the applicant's address. You or your client could be living at one address but own another property.

Part 4 – This section asks for brief particulars of the development. The word 'brief' should be noted. Do not write down a massive rambling description. If you are submitting plans for a 'single-storey kitchen extension to dwelling' then that is all is required.

If the Planning Officer needs amplification or the wording amending then he or she will contact you and say so. However, be careful. If your plans indicate a kitchen extension at the rear of the house and a porch at the front, make sure that you don't forget the porch. if you only ask for planning permission on the rear extension, then unless the case officer realizes that you have forgotten to mention the porch, you might end up only getting

(a)

Date Rec. [ ] Application No. (official use only) [ ]

TOWN AND COUNTRY PLANNING ACT 1990

Householder Planning Application Form

HALTON BOROUGH COUNCIL

USE THIS FORM FOR ANY PLANNING APPLICATION WHICH INVOLVES EXTENSIONS TO A DWELLING (INCLUDING THE ERECTION OF A GARAGE, CAR PORT, BUILDINGS, WALLS OR FENCES).

Send **THREE** copies of this form, together with **THREE** sets of plans, the appropriate certificate and fee to : Operational Director - Environmental Health & Planning Department, Environment Directorate, Halton Borough Council, Rutland House, Halton Lea, Runcorn, WA7 2GW

1 Applicant's Details (block capitals please)

Name MR & MRS J SOAP,
Address 7, SOMEWHERE ROAD, HALTON
Post Code: H1 XYZ Tel: = = =
E-Mail Address: jsoap@ - - - -

2 Applicant's Agent (If applicable)

Name ANDREW R WILLIAMS
Address 437, WARRINGTON ROAD, RAINHILL
Post Code: L35 4LL Tel: 0151 426 9660
E-Mail Address: andrew@ .... ---.

3 Full postal address of the application site (if different from the applicant's address)

7, SOMEWHERE ROAD, HALTON H1 XYZ

4 Brief description of proposed development (stating number of storeys and position in relation to the buildings)

SINGLE STOREY DINING / KITCHEN EXTENSION

5 Type and colour of external materials to be used (please specify make and colour)

a. External wall facings TO MATCH EXISTING
b. Roofing materials TO MATCH EXISTING
c. What is the maximum height of the proposed structure? 3·50 m (APPROX)
Please show on detailed drawings.

6 Access to the highway/drainage

Do you intend to:
a. Construct a new access to a highway ? YES ☐ NO ☑ VEHICULAR ☐ PEDESTRIAN ☐
b. Alter an existing access to a highway ? YES ☐ NO ☑ VEHICULAR ☐ PEDESTRIAN ☐
If YES please show this on the site plan.
c. How many parking spaces are to be provided on the site? 2
d. By what method is it proposed to drain the property/land? MAIN DRAINAGE
(If not main drainage, give full details)

7 Trees

a. Are there any trees on this site ? If YES please show them on the site plan YES ☐ NO ☑
b. Are there any trees to be felled ? If YES please show them on site plan YES ☐ NO ☑
c. Do you intend to carry out any other works to the trees on site? YES ☐ NO ☑
If YES please specify

Fig. 6.1 Householder planning application form. (Reproduced with permission from Halton Borough Council.)

(b)

8 Demolition of buildings

Do you intend to demolish any buildings, walls or fences ? YES ☐ NO ☑

If yes please show them on the detailed drawings.

9 Building Regulation Consent (most buildings need this separate consent)

Have you submitted an application for Building Regulation Consent? YES ☑ NO ☐

10 Plans

Please attach the following plans:

a. Location plan at a minimum scale of 1:1250 with site outlined in red.

b. Detailed building plans, showing existing situation and the proposed works.

Please give plan numbers and a list of information supplied:

F1/1 + ORDNANCE SURVEY SHEET (SITE BOUNDARY MARKED IN RED)

11 Fees (normally a cheque made out to Halton Borough Council in accordance with published scale fees)

Please attach the appropriate fee £ 135·00

12 Please complete

I/We apply for full planning permission for the proposals described in this application and the accompanying plans.

SIGNED Date — — -

On Behalf of MR & MRS J SOAP.

Applicants checklist

a. THREE copies of completed forms, signed and dated.

b. At least one copy of the Certificate of Ownership signed and dated.

c. THREE copies of appropriate plans and drawings. (include O.S. based location plan)

d. Appropriate fee.

PLEASE NOTE:-

You MUST complete a Certificate of ownership.

* Complete Certificate A if you are the sole owner of all the land too which the application relates.

* Complete CERTIFICATE B if you are not the sole owner of all the land to which the application relates. In this case any other owner/s must be told about your application and you do this by serving Notice No. 1 on each owner. Please note that if any part of the building works, including the foundations or roof overhang, encroaches onto your neighbour's land, you must complete Certificate B and serve Notice No.1 on the adjoining owner.

Fig. 6.1 contd.

permission for one part of your scheme and have to re-apply for the part that you have forgotten to mention (and possibly pay an additional fee).

Part 5 – This section asks for the type and colour of materials on simple house plans, and the spaces can usually be filled in as follows:

(a) Facing brickwork to match existing.
(b) Tiles to match existing.
(c) Make sure that you insert the correct height to avoid queries. I usually insert the height – followed by the word 'approx' (i.e. 3.5 m approx.).

Part 6 – Access to highway/Drainage. This section asks for access to the highways/drainage.

Tick the relevant box for 6a and 6b.

6c is there to ensure that the alterations/extensions are not going to lead to on-street parking. For instance, if you were converting a garage into living accommodation, the planners want to ensure that you have an alternative place at the property to park the car(s). Give this question some thought and make sure you mark the proposed parking bays on the plans. Too few parking spaces can be grounds for a planning refusal.

6d asks for drainage details. Main drainage is not automatic. In many rural areas, farms have a septic tank. If there is no main drainage, the planners will wish to ensure that the drainage facilities are adequate because bad drainage and sewage contamination can lead to the spread of typhoid and other diseases.

Part 7 – Trees. This section is fairly easy to answer. You must bear in mind that some trees have preservation orders on them and you cannot remove preserved trees without permission. Although it is unlikely that small trees are covered by an order, well established trees might well be. Show any tree affected by the proposal on your plan and answer the question truthfully.

But be careful.

Like most planning questions, they are there for the best of motives. At one time, developers used to destroy beautiful trees without a moments thought. However, the presence of a tree can sometimes prevent an application being successful.

An example

Two neighbours decided to build new houses in their rear gardens and I had the commission to design both houses. Before you form the view that I hate trees, this is far from the truth. I like trees. Unfortunately, as you will discover later on the book, trees can damage the foundations of houses and extensions. So like it or not, sometimes trees have to go.

In this case, I checked with the local authority and none of the trees on the two sites were covered by Tree Preservation Orders (TPOs). They were therefore fair game, as far as I was concerned.

I advised both parties (Mr Chopper and Mr Doolittle) to remove any trees that would be in the way of new development prior to making the planning application. True to his name, Mr Chopper took my advice, went down the garden like a lumberjack and removed all of his trees. Mr Doolittle contemplated his navel.

The net result was when I made the planning application for Mr Chopper, I didn't have to declare any trees to be removed. On the second application, for Mr Doolittle, I had to notify the local planners that three trees needed removing. You have probably guessed the rest.

Mr Chopper was given permission to develop, Mr Doolittle wasn't.

The moral of the story is sometimes it's wise to remove a tree before a planning application is made (as long as it doesn't have a TPO or is in a protected area like a conservation area), just so that you don't have to tick the wrong box.

Part 8 – Demolition of Buildings. You either are or are not intending to demolish something. Tick the correct box.

The demolition of an old extension is unlikely to affect the application. The removal, for instance, of an old unsightly structure could well be a benefit not only to the applicant but also to the surrounding area. However, in conservation areas (see Chapter 7) demolition of old structures may not be allowed.

Part 9 – Building Regulations. This section has been included to make sure that the applicant does not forget to apply for Building Control approval.

Part 10 – Plans. This section has been included to confirm to the Planning Officer that he has all the relevant documents to hand when he is making his decision. Sometimes, documents can go missing. In most cases, all that is required is to state the plan number on your drawing (this plan number could just be something simple such as Plan Number Al or, if you have a numbering system, something more complex such as D/ARW/LF93/JN45B).

Ordnance Survey Sheet – You will also need to enclose copies of a site-centered Ordnance Survey (OS) Sheet (see Fig. 1.1 – Site location plan – If you have not drawn an extract of the OS sheet on your main plan).

Technical catalogues/Photographs – Sometimes, submitting technical catalogues or photographs can be helpful.

Letters from a neighbour – If the Planning Officer knows that your neighbours have been consulted this can sometimes be very helpful to an application.

Other relevant information – Sometimes a personal statement can help an application.

Part 11 – Fees. When you make a planning application the local authority charge a fee for accepting the submission. If your client does not pay the planning fee, the application will not be considered.

Part 12 – This is a formal application confirming that you wish the local authority to consider your application.

Last but not least – All Planning Applications have to be accompanied by a Section 66 certificate. The most common is certificate A.

Beware. If you are building up to your boundaries (line of junction) the local authority will expect you to submit a Certificate B plus a Notice No. 1 to the neighbour (Fig. 6.2).

IN MOST CASES EITHER CERTIFICATE A

HALTON BOROUGH COUNCIL

UNDER SECTION 66 OF THE TOWN AND COUNTRY PLANNING ACT 1990

Certificate B

I hereby certify that:

1 The requisite Notice No.1 has been given to the owner(s) of the land to which the application relates at the beginning of the period of 20 days before the date of the accompanying application.

Name and address of owner MR & MRS BLOGGS
9, SOME WHERE ROAD, HALTON H1XYZ

2 None of the land to which the application relates constitutes or forms part of an agricultural holding.

SIGNED Date — —

On Behalf of MR & MRS J. SOAP.

TO BE FORWARDED TO THE OWNER OF THE LAND BY THE APPLICANT

HALTON BOROUGH COUNCIL

UNDER SECTION 66 OF THE TOWN AND COUNTRY PLANNING ACT 1990

Notice No.1

An application for Planning Permission is being made to Halton Borough Council and you are owner/part owner of the application site.

Address of application site 7, SOME WHERE ROAD
HALTON H1XYZ

Description of the proposed development SINGLE STOREY KITCHEN/DINING EXTENSION.

Name and address of applicant MR & MRS J. SOAP 7, SOMEWHERE ROAD,
HALTON

If you wish to make representations on this proposal, please do so within three weeks of receiving this notice to the:
Operational Director - Environmental Health & Planning Department, Environment Directorate, Halton Borough Council, Halton Lea, Runcorn, WA7 2GW

SIGNED Date — —

On Behalf of MR & MRS J. SOAP.

Fig. 6.2 Certificate B and Notice 1

Certificate A is used where no one other than the applicant has an interest in the land (e.g. the applicant is an owner-occupier with a normal mortgage and his or her proposals do not interfere with the neighbours). When you sign this form you are certifying the following:

(a) That the applicant owns the land.
(b) That none of the land is part of an agricultural holding.

Certificate B is used when another party has an interest in the land (or part of the land). When you sign this form you are certifying the following:

(a) That you have notified all other owners of the land.
(b) That none of the land is part of an agricultural holding.

7 Conservation areas/listed buildings/tree preservation orders

CONSERVATION AREAS

If you refer to Fig. 7.1(a)–(g) you will see a typical Conservation Area Policy. You will note that the local authority in question has made specific requirements concerning alteration works carried out in this conservation area. When dealing with work in a conservation area you must be prepared to make enquiries with the local authority regarding their policies.

In conservation areas, Permitted Development rules are sometimes suspended by an Article 4 Direction and planning permission is required for most alterations.

LISTED BUILDINGS

The effects of listing are similar to those created by work in conservation areas. When buildings are listed and graded, there is an obligation on the part of the local authority to notify the owner and occupier as soon as their house is placed on the list. It is unlikely therefore that an owner/client will be unaware that their house is listed. In general, if a property is listed it means that demolition, alteration and additions will be allowed only after proposals have been carefully examined and alterations or extensions must not deface the character of the original. The demolition or alteration of a listed building without prior permission is punishable by imprisonment and/or a heavy fine. Normally, when dealing with an alteration to a listed building a separate application in addition to planning application must be made to the local authority. I would suggest that you contact your local authority and ask for an application form for alteration to a listed building so that you can study the contents. In general, alterations to listed buildings have to be advertised in local newspapers giving details of the proposals and saying where the plans may be inspected. A notice giving similar details is usually put on the site.

TREES

Generally

The effects of tree roots and moisture extraction on foundations has been discussed elsewhere but you need to consider surrounding trees when submitting a planning application.

The law protects trees in several ways:

(a) In general the local authority have to be notified of any work to trees in a conservation area.
(b) Ditto if the tree is the subject of a tree preservation order.

(**Note**: The Forestry Commission can become involved if a large number of trees are involved.)

(a)

RAINHILL CONSERVATION AREA

Rainhill village centre is now a conservation area. This pamphlet has been prepared to inform owners and occupiers in the area of the implications of this. It describes Rainhill's history and character, the Council's policy for conservation and the effects of the extra controls introduced to safeguard the appearance of the area.

Fig. 7.1(a)–(g) Rainhill conservation area policy. (Reproduced with permission from St Helen's Borough Council.)

(b)

The History and Character of the Area

The village of Rainhill has developed over a long period of time. The name Rainhill is thought to derive from the Old English personal name of Regna or Regan. The earliest known reference is in 1190 when Richard de Eccleston granted to Alan the Clerk, his brother, the vill of Raynhull.

Until the late nineteenth century the village was an agricultural community. Some of the early recorded villagers reflect the nature of the community:

Edward Halsall, blacksmith, 1662;
Edward Whitlow, butcher, 1663;
Aeron Phythian, millwright, 1736;
Edward Parr, miller, 1713;
Thomas Hey, shoemaker, 1662;
George Smethers, stocking weaver, 1723;
Richard Glover, stonemason, 1620;
Thomas Glover, tanner, 1620;
Henry Garnett, weaver, 1635;

In addition there were the yeomen, husbandmen and labourers who worked on the farms.

The two most important events in the history of Rainhill were, first, the turnpiking of the highway from Liverpool to Warrington in the mid eighteenth century resulting in the improvement of the highway running through Rainhill from east to west and facilitating the development of the coaching service. Secondly, the construction of the Liverpool Manchester railway in the 1820's which helped its development later as a residential area. This railway is the first inter-city passenger railway not only in Britain but in the world.

In 1829 Rainhill was the venue for the locomotive trials to decide the type of engines which would operate on the railways. The engines had to traverse the level portion of the line from Rainhill Bridge towards Manchester and to complete a total run of 70 miles (the distance from Liverpool to Manchester and back) by 40 trips along the test length. Only the Rocket met the conditions and was declared the winner. These trials were of great importance in establishing the steam locomotive as the means of traction on the railways. The railway bridge which carries the Warrington Road over the railway is of particular importance. It was not possible to divert the road or railway to make a rectangular crossing and the bridge had to cross at an angle of 34°, forming a 'skew' bridge. The bridge is important as an early example, if not the first, of its kind. The station too is of interest. The original station built in 1830 was called Kendricks Cross Station (a reference to Kendricks Cross which used to stand at the junction of Warrington Road and View Road) and was to the east of its present position. There was a level crossing joining what are now Victoria Road and Tasker Terrace thus maintaining the old highway from Cronton to Eccleston. The present station was built in the 1870's and retains its Victorian character.

The conservation area is concerned with the historic core of the village which developed along Warrington Road, either side of Kendricks Cross and the area between Warrington Road and the railway. This area is the present village centre which continues to fulfil its traditional function as a local centre containing the parish church, schools, shops, services, etc.

The character of the area is formed by the linear grouping of the buildings along Warrington Road. Although none of these buildings are outstanding, grouped together they form a pleasant environment. The area is visually enclosed by the rise of Warrington Road to the west to pass over the railway, by the curving tree-lined road to the east and by the buildings and large trees fronting onto Warrington Road. This enclosure together with the scale of the buildings gives the area a human scale which is worthy of conservation. The buildings also provide a link with the past, particularly the sandstone church and the old school buildings, giving a sense of continuity and stressing the development of the village from this small original core. The red sandstone seen in the old walls and buildings in fact bears witness to one of the oldest crafts in the township. A number of quarries were being worked in the area at one time but all have now ceased. Stone from View Road Quarry was used to build St. Ann's Church.

In order to protect the character of the area the Council intend to make Article 4 Directions. These have the effect of restricting specified classes of permitted development for which planning permission is not normally required, such as alterations to surrounding walls, building small extensions, garages, porches etc., changing windows and doors. A full list is given below. It should be stressed that it is not intended to prevent alterations and improvements from being carried out, but rather to ensure that they are carefully designed to be in character with the original. The Planning Section of the Council's Technical Services Department will give you advice prior to you making a formal application for planning permission should you be contemplating works on your property.

There is great scope for improvement within the conservation area. Most of these are of a minor, environmental nature, such as the landscaping and tidying up of vacant sites, landscaping of parking areas, painting and tidying up of some buildings, attention to walls and fences. Some of this work is the responsibility of the Local Authority, some is the responsibility of individual owners or occupiers. Proposals for the improvement of the area will be submitted to a public meeting to be held in Rainhill to enable all interested people and organisations to discuss the proposals.

Milestone on Bridge, Warrington Road (1829).

Fig. 7.1 contd.

(c)

Article 4 Directions

The following classes of permitted development are to be restricted by the making of Article 4 Directions. Planning permission is therefore required for these works.

Class I – Development within the curtilage of a dwelling house.

(a) The enlargement, improvement or other alteration to a dwelling house.

(b) The erection or construction of a porch.

(c) The erection, construction or placing, and the maintenance, improvement or other alteration, within the curtilage of a dwelling house, of any building or enclosure required for a purpose incidental to the enjoyment of the dwelling house.

(d) The construction of a hardstanding for vehicles.

(e) The erection or placing of an oil storage tank for domestic heating.

Class II – Sundry minor operations

(a) The erection of construction of gates, fences, walls or other means of enclosure.

(b) The formation, laying out and construction of a means of access to a highway.

(c) The painting of the exterior of any building.

Class VIII – Development for industrial purposes

(a) Development of the following descriptions carried out by an industrial undertaker:

(i) the provision, re-arrangement or replacement of private ways;

(ii) the provision or re-arrangement of sewers, mains, pipes, cables or other apparatus;

(iii) the installation or erection, by way of addition or replacement, of plant or machinery;

(iv) the extension or alteration of buildings.

(b) The deposit of an industrial undertaker of waste material or refuse resulting from an industrial process.

Class IX – Repairs to unadopted streets and private ways.

The carrying out of works required for the maintenance or improvement of an unadopted street or private way.

Class XIII – Development by Local Authorities

(a) The erection or construction and the maintenance, improvement or other alteration by a local authority of:

(i) such small ancillary buildings, works and equipment as are required on land belonging to or maintained by them;

(ii) lamp standards, information kiosks, passenger shelters, public shelters and seats, telephone boxes, fire alarms, public drinking fountains, horse troughs, refuse bins or buckets, barriers for the control of persons waiting to enter public vehicles, and such similar structures or works as may be required in connection with the operation of any public service administered by them.

(b) The deposit by a local authority of work material or refuse on any land.

Class XIV – Development by local highway authorities

The carrying out by a local highway authority of any works required for or incidental to the maintenance or improvement of existing highways.

Class XVIII – Development by statutory undertakers

All development by statutory undertakers.

Fig. 7.1 contd.

(d)

Warrington Road.

Policy

Conservation implies a long term view. It implies a responsibility to leave our historic environments at least as good as we found them. The need for Conservation can be argued in several ways: the links with our past provide us with a perspective against which we can evaluate our present actions; continuity in the environment enables us to have a sense of identity with familiar surroundings and this perhaps contributes to emotional and community stability; on a practical level conservation enables the careful management of limited resources.

Conservation cannot be considered in isolation. It should be seen in the context of other Borough Council functions. Designating an area as a conservation area does not mean that it will be preserved in its entirety as a museum piece. This would be unrealistic and would not take into account the other responsibilities of the Borough Council nor the social and economic forces over which the Borough Council has no control. An area must adapt to meet changing requirements and must be provided with those elements which allow it to continue to function. However, in designating a conservation area the Borough Council recognises the environmental quality of that area, and the importance of retaining this quality will be reflected in any future planning proposals and decisions. The Council will also take specific steps to protect, and where necessary improve, the environment within designated conservation areas. These steps are described below.

1 Development Control

The following general principles will be adopted for dealing with planning applications in conservation areas.

(a) Outline applications for planning permission will not normally be considered without sufficient details of the siting and design of a proposed building to show the proposal in the context of its surroundings.

(b) A high standard of design for new buildings and for the alteration or restoration of existing buildings will be expected. Any new building should be so designed as to harmonise in form, scale and materials with the area as a whole and with its immediate surroundings in particular. Care should be taken that the siting of proposed buildings is in sympathy with the pattern of existing frontages.

(c) Attention will be given to the proper planting and maintenance of trees. Developers will be encouraged to plant additional trees where appropriate as part of their development proposals as well as carrying out other landscaping work.

(d) It is important that materials used in a conservation area are sympathetic to that area and reflect its character. The use of cobbles, granite setts, brick pavers, stone flags etc. will be encouraged where appropriate and the Borough Council will keep a store of such materials. Careful attention will also be paid to the design and siting of street furniture and steps will be taken where possible to limit the intrusion of wires, posts, pipes etc.

2 Policies specific to individual Conservation Areas

Where necessary and when the opportunity arises detailed policies will be introduced for conservation and improvement in specific conservation areas. The introduction of these policies will depend on the availability of financial resources. Any proposals will be submitted for consideration to a public meeting in the areas to which they relate.

3 Residents in Conservation Areas

The Borough Council recognises that the success of any conservation policy depends upon the co-operation and the enthusiasm of residents, traders and property owners in the conservation areas. Even if the Borough Council was not strictly limited in manpower and in financial resources it would be vitally important that residents accept some of the responsibility for the appearance of their local environment. Much work has been done in the past by volunteer groups – clearing up of waste land, tree planting, landscaping, clearing of canals and ponds, restoration of buildings. The Borough Council will encourage any local initiative of this kind and give as much assistance as possible both in the organisation and implementation of such projects.

4 Conservation Advisory Group

The Council has established a Conservation Advisory Group consisting of representatives from Parish Councils, local and national amenity societies, local Chamber of Trade etc. The group is able to provide specialist and local knowledge and acts as a liaison body between the public and the Borough Council on Conservation matters.

The group may be contacted through its Chairman, c/o The Town Clerk, Town Hall, St. Helens.

Fig. 7.1 contd.

Fig. 7.1 contd.

(f)

Legislation

This Section summarises the relevant powers and duties of the Borough Council with regard to Conservation.

1 Designation of Conservation Areas

The designation of conservation areas is made under Section 277 of the 1971 Town and Country Planning Act as re-enacted by Section 1 of the Town and Country Amenities Act, 1974 which requires the Local Planning Authority to determine which parts of their area are of special architectural and historic interest, the character or appearance of which it is desirable to preserve or enhance and designate such areas as conservation areas.

In order to protect the character and appearance of conservation areas it is necessary to bring development under tighter control. The powers to do this, made available through various Acts of Parliament, are outlined below.

2 Demolition of Buildings

A building in a conservation area cannot be demolished without the consent of the Local Authority. An application to demolish a building may be made as a separate application or as part of an application for planning permission.
(Town and Country Planning Act, 1971, Section 55).

3 Urgent Repair of Unoccupied Buildings

If it appears to the Local Authority that any works are urgently required for the preservation of an unoccupied building in a conservation area and that it is important to preserve the building for the purpose of maintaining the character or appearance of the conservation area, then they may execute the works after giving the owner of the building not less than seven days' notice, in writing, of their intention to do so.

The Local Authority may give notice to the owner of the building requiring him to pay the expenses of any works executed.
(Town and Country Amenities Act, 1974, Section 5).

Fig. 7.1 contd.

(g)

4 Grants and Loans

(The term 'grant' in this section can be taken to include loans).

Grants are available in certain circumstances from both central government funds and from local authorities. They are always at the discretion of the body giving them and are naturally restricted by the amount of money available at a given time.

Exchequer Grants

The Secretary of State for the Environment has power to make grants for the repair and maintenance of buildings that are of outstanding architectural or historic interest. Comparatively few buildings qualify and so the scope for these grants is limited. (Historic Buildings and Ancient Monuments Act; 1935, Section 4). (Civic Amenities Act, 1967 Section 4).

Local Authority Grants

Local authorities have a wider scope. They can make grants for any building of architectural or historic interest and are not restricted to outstanding buildings. Again grants are payable for works of repair and maintenance. Grants are not payable for alterations or additions.
(Local Authorities (Historic Buildings) Act, 1962.
(Town and Country Planning Act, 1968, Section 58).

You may also be able to get a home improvement grant for improving or converting a building which is to be used as a dwelling.
(Housing Act, 1974).

5 Publicity for Planning Applications

Notice of any application for planning permission that would affect the appearance or the character of a conservation area is published in a local paper, stating where copies of the application and plans may be inspected by the public. The Local Authority will also, for not less than seven days, display on or near the land, a notice indicating the nature of the development in question. The application is not determined for 21 days from the date of publication of the notice. During this period the public may make representations to the Local Authority and any such representations are taken into account before the application is determined. This procedure may also apply to applications on land adjacent to a conservation area where the proposed development would affect the character of the conservation area. (Town and Country Planning Act, 1971, Section 28).

6 Protection of Trees

Anyone wishing to carry out work on a tree or to fell a tree in a conservation area must give the District Council six weeks notice of his intention. During this time the Council may serve a tree preservation order on the tree. A public register of such notices is to be kept and the notices are valid for two years. Heavy fines are now imposed on any person who, in contravention of a tree preservation order, cuts down, uproots or wilfully damages a tree in such a manner as to be likely to destroy it.
(Town and Country Amenities Act, 1974, Section 8).

7 Control of Advertisements

The Local Authority can control the display of all advertisements in a conservation area.
(Town and Country Amenities Act, 1974, Section 3).

8 Town and Country Planning General Development Order, 1973

The Local Authority has the power to make the following directions subject to Ministerial confirmation for the protection of conservation areas:

(a) An Article 4 Direction. If it becomes necessary this direction allows the Local Authority to bring any class of "permitted" development (i.e. certain small scale alterations or additions) under specific development control in the whole or any part of a conservation area.

(b) An Article 5 Direction. Where outline planning permission ought not to be considered separately from questions of siting or design or external appearance of the building, access, or landscaping of the site, the Local Authority may within one month of the application notify the applicant that they are "unable to entertain it unless further details are submitted". The Local Authority must specify what further information they need for arriving at a decision. The applicant may either furnish the information or appeal to the Department of the Environment within six months of the notice.

Published by G. K. Perks, C.Eng., M.I.C.E., M.I.Mun.E.,

Fig. 7.1 contd.

Trees in conservation areas

The general principle is that you need to write to the local authority if you intend to carry out work or remove a tree in a conservation area.

Tree preservation orders

A local authority may make a tree preservation order relating to any tree, group of trees or a belt of woodland, and these may be in fields, gardens or building sites. Hedges are not covered, but large trees growing in hedges could be. You cannot interfere with a preserved tree without express permission. This is one of the reasons why the planning application form asks if any trees are to be felled. Once the Planning Officer knows that a tree is going to be removed, he or she will check to ensure that the tree concerned has no order on it.

8 Other design aspects for consideration

GENERALLY

Despite what some of the planning guidelines say about extensions being made to look like subordinate structures, I believe that unless the requirements of the planning authority force you down the 'subordinate route' the objective when designing an extension, if at all possible, should be to make it look like part of the original structure and not an afterthought.

BRICKWORK DIMENSIONS

Figures 8.1 and 8.2 are extracts from Ibstock's Technical Notes and are based upon coordinating brick sizes of 225 × 112.5 × 75 m which includes 10 mm joints. When designing an extension to an existing property, the designer should wherever possible attempt to keep brick cutting to a minimum. Obviously in older properties the designer should do his or her best to provide a match as described below.

MATCHING MATERIALS

One of the problems that extensions create is that of matching new materials with the existing. For the most part, the matching of materials on very small domestic extensions does not come within the remit of the surveyor because he or she has carried out his or her function merely by stating on the plans that the 'facing bricks shall match the existing' or that 'the new roof tiles shall match the existing'. The client then passes this problem over to the builder.

But what if you are engaged to supervise the works, or the extension is for yourself? Then it is necessary to know how to find a good match.

IBSTOCK

TECHNICAL NOTES: CLAY BRICKS & BRICKWORK DIMENSIONS

Brickwork Dimensions

In the production of design and construction drawings for brickwork, the designer should think *Bricks* and minimise cutting by using brick dimensions wherever possible. An awareness that the architect understands the craftsman's problems, such as unnecessary, time consuming and costly cutting, encourages interest in the job and a better result. Brickwork dimensions tables are intended as an aid to architects and designers in the preparation of design and construction drawings.

Notes

(i) When the bricks to be used in piers or small width panels of up to 4 bricks are all either above or below *work size*, adjustment of the pier or brick panel width should be considered to avoid unusually large or small width perpend joints.

(ii) It is helpful to the site to indicate on construction drawings the number of bricks which make up the dimension.

(iii) Further information on the placement of expansion joints will be found in Ibstock Technical Note – *Brickwork, Designing for Movement* and in BS 5628 Part III 1985 and *Code of Practice for Use of Masonry*.

(iv) Based upon bricks produced to BS3921 with *co-ordinating sizes* of 225 x 112.5 x 75 which includes 10mm joints.

Vertical brickwork courses dimensions tables

using 65mm bricks and 10mm joints

COURSES		COURSES		COURSES		COURSES	
1	75	31	2325	61	4575	91	6825
2	150	32	2400	62	4650	92	6900
3	225	33	2475	63	4725	93	6975
4	300	34	2550	64	4800	94	7050
5	375	35	2625	65	4875	95	7125
6	450	36	2700	66	4950	96	7200
7	525	37	2775	67	5025	97	7275
8	600	38	2850	68	5100	98	7350
9	675	39	2925	69	5175	99	7425
10	750	40	3000	70	5250	100	7500
11	825	41	3075	71	5325	101	7575
12	900	42	3150	72	5400	102	7650
13	975	43	3225	73	5475	103	7725
14	1050	44	3300	74	5550	104	7800
15	1125	45	3375	75	5625	105	7875
16	1200	46	3450	76	5700	106	7950
17	1275	47	3525	77	5775	107	8025
18	1350	48	3600	78	5850	108	8100
19	1425	49	3675	79	5925	109	8175
20	1500	50	3750	80	6000	110	8250
21	1575	51	3825	81	6075	111	8325
22	1650	52	3900	82	6150	112	8400
23	1725	53	3975	83	6225	113	8475
24	1800	54	4050	84	6300	114	8550
25	1875	55	4125	85	6375	115	8625
26	1950	56	4200	86	6450	116	8700
27	2025	57	4275	87	6525	117	8775
28	2100	58	4350	88	6600	118	8850
29	2175	59	4425	89	6675	119	8925
30	2250	60	4500	90	6750	120	9000

Fig. 8.1 Vertical brickwork dimensions. (Reproduced with permission from Ibstock Building Products Ltd.)

TECHNICAL NOTES: CLAY BRICKS & BRICKWORK DIMENSIONS — IBSTOCK

Horizontal brickwork dimensions tables

using 215 x 102.5 x 65mm bricks and 10mm joints

Number of bricks	Openings	Panels with opposite return ends (co-ordinating size)	Piers or panels between openings
½	122.5	112.5	102.5
1	235	225	215
1½	347.5	337.5	327.5
2	460	450	440
2½	572.5	562.5	552.5
3	685	675	665
3½	797.5	787.5	777.5
4	910	900	890
4½	1022.5	1012.5	1002.5
5	1135	1125	1115
5½	1247.5	1237.5	1227.5
6	1360	1350	1340
6½	1472.5	1462.5	1452.5
7	1585	1575	1565
7½	1697.5	1687.5	1677.5
8	1810	1800	1790
8½	1922.5	1912.5	1902.5
9	2035	2025	2015
9½	2147.5	2137.5	2127.5
10	2260	2250	2240
10½	2372.5	2362.5	2352.5
11	2485	2475	2465
11½	2597.5	2587.5	2577.5
12	2710	2700	2690
12½	2822.5	2812.5	2802.5
13	2935	2925	2915
13½	3047.5	3037.5	3027.5
14	3160	3150	3140
14½	3272.5	3262.5	3252.5
15	3385	3375	3365
15½	3497.5	3487.5	3477.5
16	3610	3600	3590
16½	3722.5	3712.5	3702.5
17	3835	3825	3815
17½	3947.5	3937.5	3927.5
18	4060	4050	4040
18½	4172.5	4162.5	4152.5
19	4285	4275	4265
19½	4397.5	4387.5	4377.5
20	4510	4500	4490
20½	4622.5	4612.5	4602.5
21	4735	4725	4715
21½	4847.5	4837.5	4827.5
22	4960	4950	4940
22½	5072.5	5062.5	5052.5
23	5185	5175	5165
23½	5297.5	5287.5	5277.5
24	5410	5400	5390
24½	5522.5	5512.5	5502.5
25	5635	5625	5615
25½	5747.5	5737.5	5727.5
26	5860	5850	5840
26½	5972.5	5962.5	5952.5
Consider expansion joint at 6mm centres in parapet and free standing walls			
27	6085	6075	6065
27½	6197.5	6187.5	6177.5
28	6310	6300	6290
28½	6422.5	6412.5	6402.5
29	6535	6525	6515
29½	6647.5	6637.5	6627.5
30	6760	6750	6740
30½	6872.5	6862.5	6852.5
31	6985	6975	6965
31½	7097.5	7087.5	7077.5
32	7210	7200	7190
32½	7322.5	7312.5	7302.5
33	7435	7425	7415
33½	7547.5	7537.5	7527.5
34	7660	7650	7640
34½	7772.5	7762.5	7752.5
35	7885	7875	7865
35½	7997.5	7987.5	7977.5
36	8110	8100	8090
36½	8222.5	8212.5	8202.5
37	8335	8325	8315
37½	8447.5	8437.5	8427.5
38	8560	8550	8540
38½	8672.5	8662.5	8652.5
39	8785	8775	8765
39½	8897.5	8887.5	8877.5
40	9010	9000	8990
40½	9122.5	9112.5	9102.5
41	9235	9225	9215
41½	9347.5	9337.5	9327.5
42	9460	9450	9440
42½	9572.5	9562.5	9552.5
43	9685	9675	9665
43½	9797.5	9787.5	9777.5
44	9910	9900	9890
44½	10022.5	10012.5	10002.5
45	10135	10125	10115
45½	10247.5	10237.5	10227.5
46	10360	10350	10340
46½	10472.5	10462.5	10452.5
47	10585	10575	10565
47½	10697.5	10687.5	10677.5
48	10810	10800	10790
48½	10922.5	10912.5	10902.5
49	11035	11025	11015
49½	11147.5	11137.5	11127.5
50	11260	11250	11240
50½	11372.5	11362.5	11352.5
51	11485	11475	11465
51½	11597.5	11587.5	11577.5
52	11710	11700	11690
52½	11822.5	11812.5	11802.5
53	11935	11925	11915
Consider expansion joint at 12mm centres in all cases			
53½	12047.5	12037.5	12027.5
54	12160	12150	12140
54½	12272.5	12262.5	12252.5
55	12385	12375	12365
55½	12497.5	12487.5	12477.5
56	12610	12600	12590
56½	12722.5	12712.5	12702.5
57	12835	12825	12815
57½	12947.5	12937.5	12927.5
58	13060	13050	13040
58½	13172.5	13162.5	13152.5
59	13285	13275	13265
59½	13397.5	13387.5	13377.5
60	13510	13500	13490
60½	13622.5	13612.5	13602.5
61	13735	13725	13715
61½	13847.5	13837.5	13827.5
62	13960	13950	13940
62½	14072.5	14062.5	14052.5
63	14185	14175	14165
63½	14297.5	14287.5	14277.5
64	14410	14400	14390
64½	14522.5	14512.5	14502.5
65	14635	14625	14615
65½	14747.5	14737.5	14727.5
66	14860	14850	14840
66½	14972.5	14962.5	14952.5
67	15085	15075	15065
67½	15197.5	15187.5	15177.5
68	15310	15300	15290
68½	15422.5	15412.5	15402.5
69	15535	15525	15515
69½	15647.5	15637.5	15627.5
70	15760	15750	15740
70½	15872.5	15862.5	15852.5
71	15985	15975	15965
71½	16097.5	16087.5	16077.5
72	16210	16200	16190
72½	16322.5	16312.5	16302.5
73	16435	16425	16415
73½	16547.5	16537.5	16527.5
74	16660	16650	16640
74½	16772.5	16762.5	16752.5
75	16885	16875	16865
75½	16997.5	16987.5	16977.5
76	17110	17100	17090
76½	17222.5	17212.5	17202.5
77	17335	17325	17315
77½	17447.5	17437.5	17427.5
78	17560	17550	17540
78½	17672.5	17662.5	17652.5
79	17785	17775	17765
79½	17897.5	17887.5	17877.5
80	18010	18000	17990

Fig. 8.2 Horizontal brickwork dimensions. (Reproduced with permission from Ibstock Building Products Ltd.)

There are several problems concerning matching brickwork and other materials with the existing. I have nominated a few for consideration as follows:

(a) Unless the brick/roof tile is a local product or in fairly wide use in your area, it is sometimes difficult to identify the original brick/tile. The identification process can be made even more difficult because

  (i) Colours fade and become stained over the years so even if it is possible to find the exact brick/tile match, the result may still be unsatisfactory.
  (ii) Bricks/tiles made in different batches vary in colour.
  (iii) Successive governments have used the building industry as an economic lever. Lack of demand during periods of 'belt tightening' has forced many manufacturers out of business. If a rival does not move in and buy up the rights to produce the range, then some products just simply become unavailable.
  (iv) Standardization makes a product obsolete. This has happened with bricks. At one time all sorts of brick sizes were available (2 in, 3 in etc.).

(b) In order to make the brickwork match as close as possible, brick bond patterns, toothing to the existing and colour of mortars used also need consideration.

POSSIBLE SOLUTIONS

Matching tiles/slates

When dealing with a single-storey extension, the problem of matching is not quite so vital as when extending an existing roof. No doubt in your travels you will have seen examples of side extensions which have extended an existing roof and has left a visible junction between old and new. It is far better practice to salvage tiles/ slates from one elevation and re-use them on the most prominent elevation to ensure that the roof covering is uniform. New tiles/slates should then be used on a less prominent elevation. Where dealing with, say, a single-storey extension where there are no abutting tiles, it is difficult to see both the older upper roof and the newer lower roof, unless you stand a long way back, and it is usually possible to approximately match the tiles in colour and the finished work not to look unsightly.

Matching bricks

If it is possible to get a good sample of an existing brick, then take it to a 'brick library'. You laugh . . . but there are such things. Whilst they are not common, large builder's merchants and specialist brick suppliers and some local authorities do have 'libraries'. Larger builder's merchants and suppliers are usually prepared to let you have sample bricks to take home/or to show to your client. It is also possible to inspect sample tiles in the same way.

Brick size

In days gone by, bricks came in a variety of sizes (2 in, 3 in etc.). It is therefore essential to check the sizes of bricks whilst on site. (Bricks should be measured as size on face plus one mortar joint on the length and height.) What chance have you got of matching the new work to the old if the old walls are in obsolete three-inch bricks? The answer is none unless you can find a brick that gives a reasonable match in your brick library or can obtain 'secondhand' bricks.

Secondhand bricks and other salvaged materials

There are some suppliers that specialize in obtaining, cleaning and reselling secondhand bricks and other salvaged materials. If and when the need arises, appropriate enquiries should be made concerning the availability of recycled products. Correct use of secondhand materials can overcome the problem of matching old to new. However, when using secondhand bricks in particular, it would be worth checking with the proposed supplier to see if they have been dipped or treated in any way. Untreated secondhand bricks can introduce problems, such as dry rot, into the property to be extended.

Matching mortar

It is possible to obtain premixed coloured mortar; several firms manufacture it (e.g. Tilcon, RMC). The mortar comes in a wide variety of shades and can usually be made to imitate the mortar colouring of the existing house. Most mortar suppliers are prepared to supply a set of small samples which give an indication of every colour that they make. Once a near match has been established, these companies will then usually supply a small pack of mortar so that a 'sample panel' can be constructed for client's approval. The 'rough stuff' that is supplied, however, has to be mixed with cement before use. Ask the supplier to advise on the amount of cement required.

Matching bond

There are several types of bond for brickwork (Figs 8.3 to 8.6). As cavity walls are now the most common form of construction, stretcher bond is normally used in cavity walls. But what if the existing house wall is solid

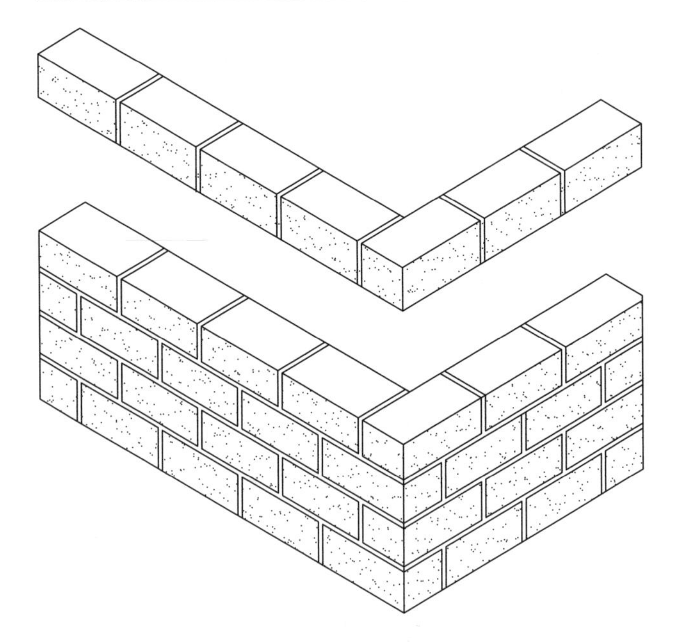

Fig. 8.3 Stretcher bond.

construction and built in English bond or Flemish bond? The way around this problem is to use 'snap headers' on the outer skin. In other words your external cavity wall will still only be half a brick thick (112 mm), but the bricklayer will chop the headers in half to imitate the old brickwork bonding that you need to match.

Vertical abutments

Where new walls abut old walls the brickwork should be cut, toothed and bonded to the existing, in other words, pockets cut in the old brickwork and the new bricks let in. If you refer to Fig. 8.7, you will note that the cutting and toothing can remain permanently on display unless carried

External appearance
shown as alternate
layers of headers
and stretchers

Fig. 8.4 English Bond.

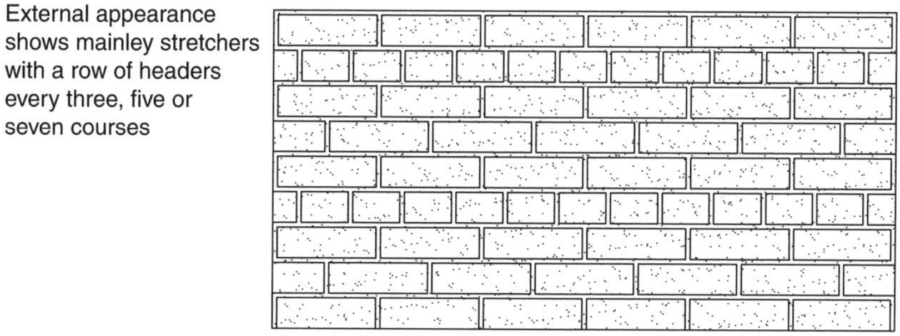

External appearance
shows mainley stretchers
with a row of headers
every three, five or
seven courses

Fig. 8.5 English garden wall bond.

out with care. Figure 8.8 shows a way to avoid the cutting, toothing and bonding showing at corners by setting the brickwork back by half a course. If you do not wish to use the solution shown in Fig. 8.8, the only way of totally avoiding cutting, toothing and bonding is by using patent jointing strips such as 'Furfix' (or other similar system) which is bolted to the existing wall and has steel 'teeth' that bed into the mortar joints of the new

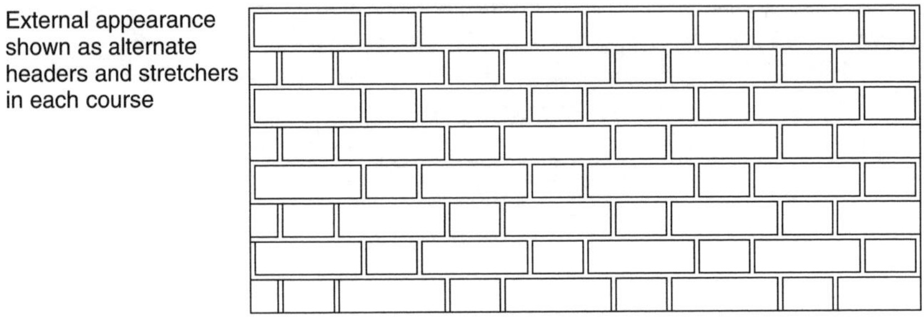

External appearance shown as alternate headers and stretchers in each course

Fig. 8.6 Flemish bond.

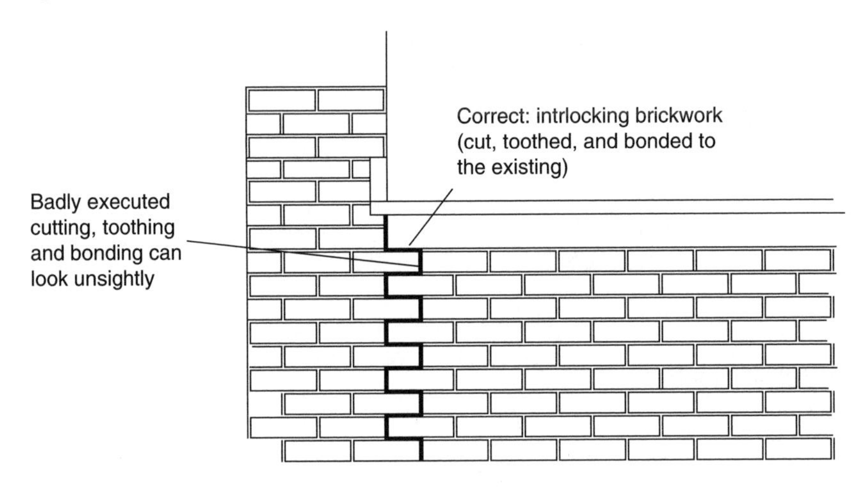

Fig. 8.7 Vertical abutments: cut, tooth and bond.

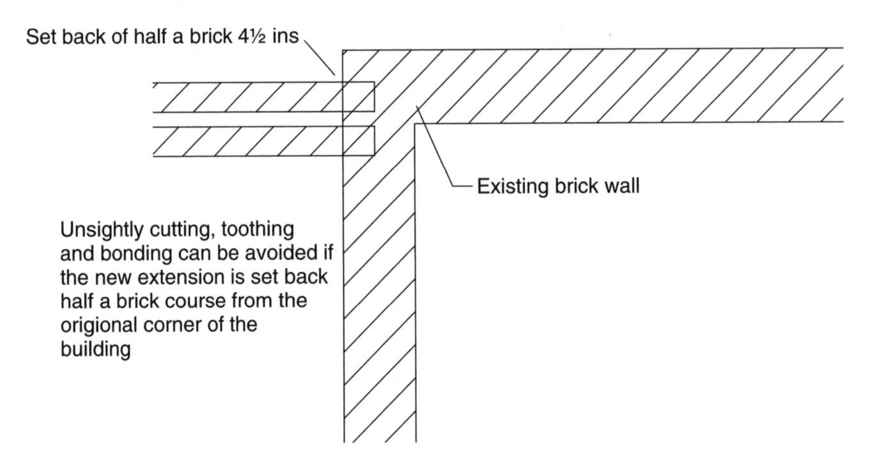

Fig. 8.8 Vertical abutments: see back at corner.

brickwork. This modern approach to jointing new to old has its advantages, but the manufacturer's instructions should be followed when using any product such as this, in order to provide a watertight joint. On the down side, the problem with using 'Furfix' or similar is that a vertical joint can be seen running down the wall face (Fig. 8.9). As this vertical jointing could look identical to 'butt jointing' (which is bad building practice), the form of construction could mislead another surveyor when visiting the house to carry out a building survey for a future buyer.

Render finishes

Where a house is totally rendered or a combination of brick and render, try to line up with the original coursing. If the old render terminates in a bellmouth (see Fig. 12.1), then put one on the new render to match in.

Lining up fascia boards and windows

Where possible, line up all horizontals. There is nothing worse than a house that has window heads at irregular heights in relation to one another. Fascia

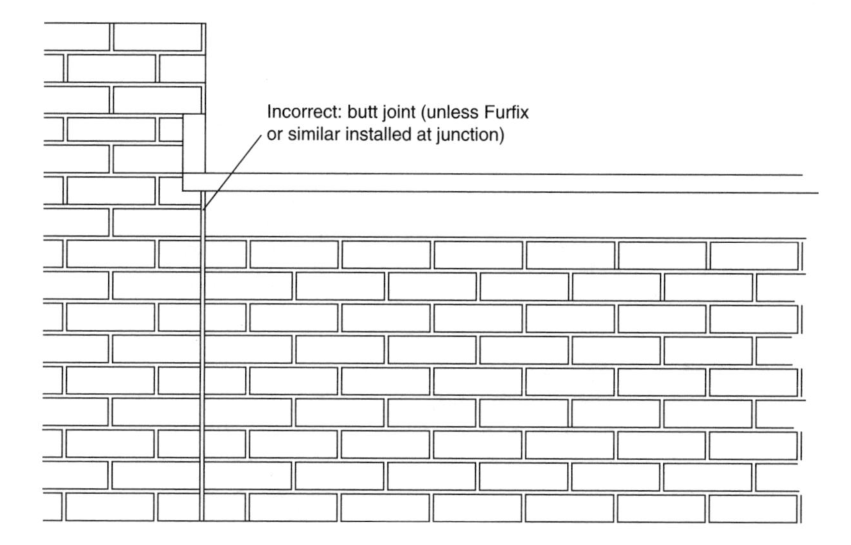

Fig. 8.9 Vertical abutments: butt joint.

boards should also be lined around if possible because it will help to create the illusion that the extension has always been there. The continuous lines help to 'weld' the extended property together.

SITE OBSTRUCTIONS

Although I will be discussing site visits in more detail later in the book, whether you are the owner or an outside consultant, you should be on the lookout for site obstructions. Here are a few examples of typical obstructions to the proposed development.

Trees

I have already made some comment on the effects that trees can have on a planning application in Chapter 6. But dealing with 'tree huggers' is not the end of the story. There are other problems caused by having trees on a site.

On page 13 of Approved Document C, there is a map showing the areas of shrinkable clay in England and Wales. As their name implies, shrinkable clays can change significantly in volume when they loose or gain water. Trees need water to survive and they can damage buildings by extracting large quantities of moisture from surrounding soils. This problem becomes more acute in a drought.

Figure 8.10 depicts a large tree growing close to a house. You will note that below ground, the roots are spreading out and can endanger the stability of the property by reducing the ground support to the foundations.

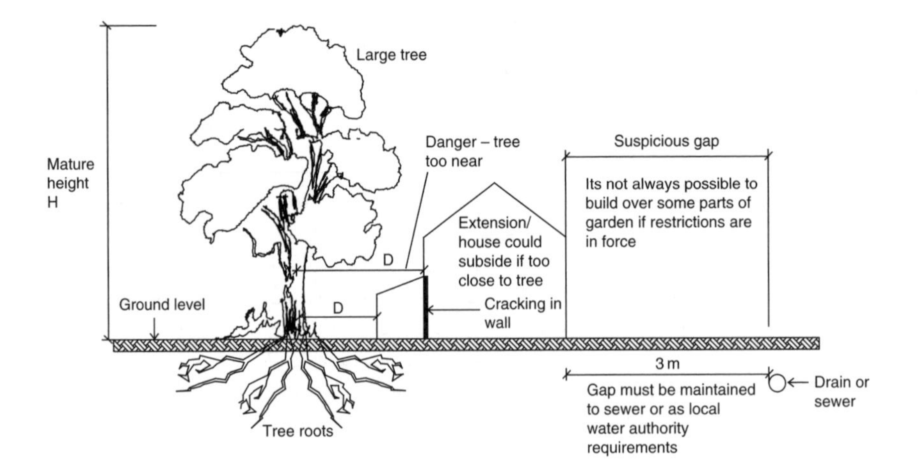

Fig. 8.10 Trees and other obstructions.

When carrying out a survey on a property, if you believe that a tree growing near a house could cause future problems, the solution is to consult NHBC Standards. This publication is designed to be used by NHBC registered builders. There is, however, no reason why the basic principals cannot be applied to home extensions.

NHBC Standards covers the subject of building near trees in a highly scientific way and provided details of likely water demand for a wide variety of trees. The schedules also provide likely heights of different types of trees and recommended foundation depths. For example, using their rules, if you were dealing with a poplar tree (anticipated height stated as 28 m) growing on a high shrinkable soil, the distance from the centre of the tree to the house has to be at least 35 m before normal footings 1 m deep can be used. At 14 m the depth has to be 2.50 m. Closer than that the foundations need to be engineer designed and special precautions must be made to protect the foundations against excessive root damage.

If a house (or extension) is built too close to tree without taking precautions, it is virtually inevitable that the property will eventually sustain damage.

As example of the sort of damage that trees can cause, I would cite a typical situation that I discovered when carrying out a homebuyer report on a detached bungalow. From the moment that I pulled up outside the house, I guessed there was a tree problem because a line of large sycamores on the rear boundary dominated the whole building.

As the trees were very close to the bungalow around 4 m away, the roots had damaged some foundations and cracks had opened up on the rear walls. The trees had also damaged a concrete garage that the owners had constructed a couple of decades before.

Obviously, when the garage had been built, the trees had only been small. But trees grow. Eventually, the trunks had expanded and had finally reached the back of the garage. As the trees continued growing, the pressure had started to push the garage off its foundations and had moved it forward 150 mm (6 inches).

The trees had also caused a number of minor defects to arise. As the tree canopy covered most of the rear roof of the bungalow, the gutters were clogged with leaves and rear tiled roof and rear flat roofs were covered in thick moss creating a constant maintenance problem.

Sewers and buried mains cables

On some housing estates you may note apparently unaccountably large gaps between your house/prospective client's property and the next house, or that you or your client has an unusually large garden. Beware! Ask questions! Does anyone know of any sewer or buried cables in the vicinity? Builders never leave large strips of land undeveloped out of the goodness of their hearts. Prime building land costs a great deal of money and they will naturally try to build as many houses on a plot as they can. If you/your prospective client thinks that there might be a main sewer or electric cable in the vicinity . . . check! (Fig. 8.10). In the first instance it would be advisable to contact the Main Drainage Section at your local council. (They may have a different title at your council but do not be put off – you want to speak to the person/section that deals with sewers.)

Note: I am using the term 'sewer' as opposed to 'drain' quite deliberately. Sewers are larger than the normal 100/150 mm drainage that serves the house itself and are normally owned and maintained by the council or the water company in your area. In my experience, most sewers owned by the water company have restrictions placed upon them, and property owners are not normally allowed to build near to a sewer. In my area, the general rule is that no permanent structure shall be built closer than 3 m from the edge of a sewer, but in one recent plan that I prepared a 4 m gap was required. If you don't make some enquiries before you start preparing plans then both you and your client could be in for a disappointment. In the worst case, you could find that your plans will be rejected totally and that no amount of amendment will make your proposals acceptable. If the obstruction just turns out to be a deep drain, then the only problem is ensuring that the foundations of the house will not subside into it in the event of a failure. This can be resolved by concrete encasing of the drain or by putting in deeper foundations to the extension. This aspect of design is covered in later chapters. Large electric cables installed by the local electricity company can also cause similar problems to sewers. If you believe that there might be a high voltage cable, water main, gas main or any other obstruction buried on the site, you can find out the truth by contacting your local electricity, gas or water companies. Normally, they will mark up a plan of their services and send it to you, without charge. The only snag is that large red disclaimer that they have stamped on the plan saying that the service position is only approximate to plus or minus six feet. If the service is very close to the proposal, you could still have a problem.

The uncooperative next door neighbour

With most small extensions, the owner of the property to be extended wishes to maximize his or her benefits by building up to the boundary of his or her property. Normally neighbours will not object to this because they will be on friendly terms with either you or your client and will accept that the neighbours desire to better themselves as a reasonable aspiration. Unfortunately this is not always the case. There are two simple solutions to the problem of the awkward neighbour. The first is shown in Fig. 8.11(a). Here the new extension is set back 150 mm which means that the footing underground will not go under the neighbour's garden (foundations normally have a 150 mm spread). The second solution is to use a tied footing as detailed in Fig. 8.11(b). As the toe of the tied footing does not need spread, the external wall face of the extension can be built on the boundary. Admittedly, the bricklayer building the wall adjacent to the neighbour may need to work 'overhand' to construct the wall on the

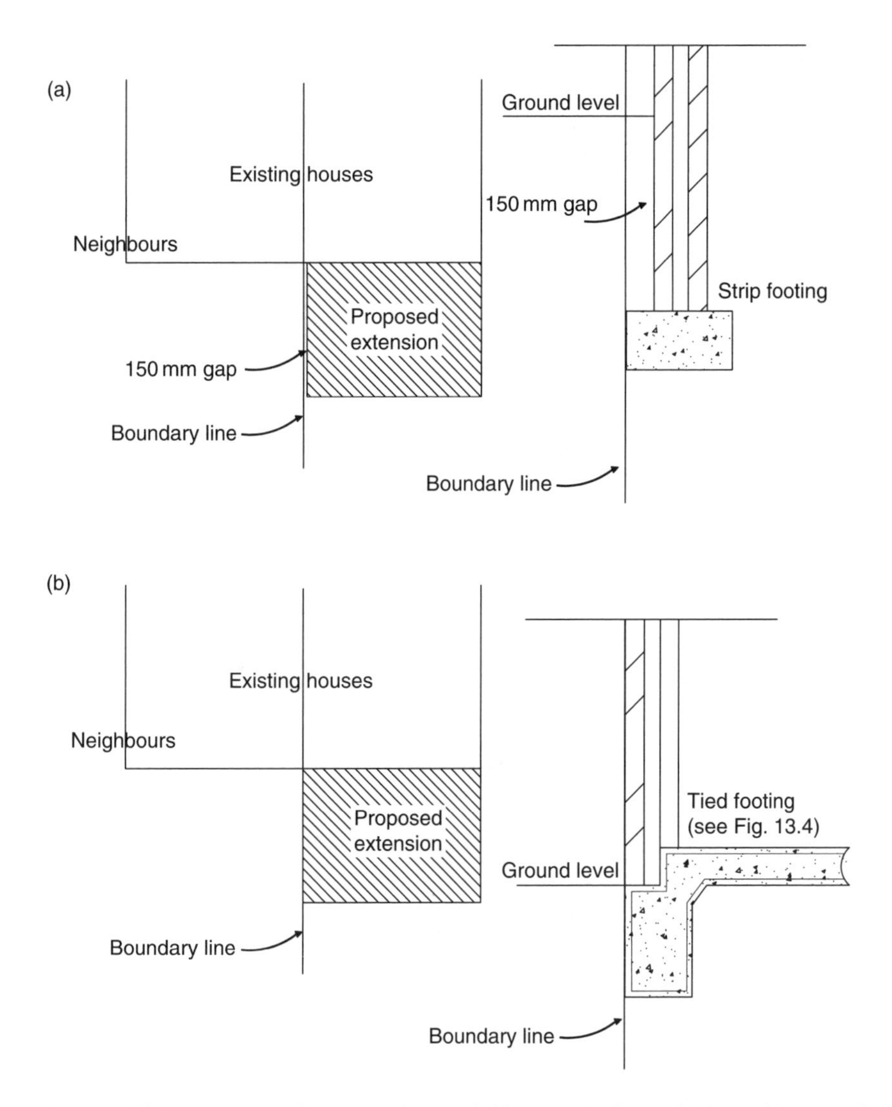

Fig. 8.11 The uncooperative next door neighbour: (a) first solution; (b) second solution.

boundary, but this does not normally prevent the extension being built. (Tied footings are dealt with in greater detail in later chapters.)

Privacy distances . . . parallel windows

As discussed before, most local authorities now have rules regarding 'overlooking'. Where there are two parallel blocks of houses (Fig. 8.12), most Planning Officers will be concerned with maintaining reasonable privacy distances between windows of neighbouring houses. The reason is obvious. It would be very distasteful if one neighbour were allowed to construct a structure very close to, say, the main living room or bedroom window of another dwelling and thereby deprive the second party of his or her privacy. Every council has its own policies on such matters and you should enquire what they are.

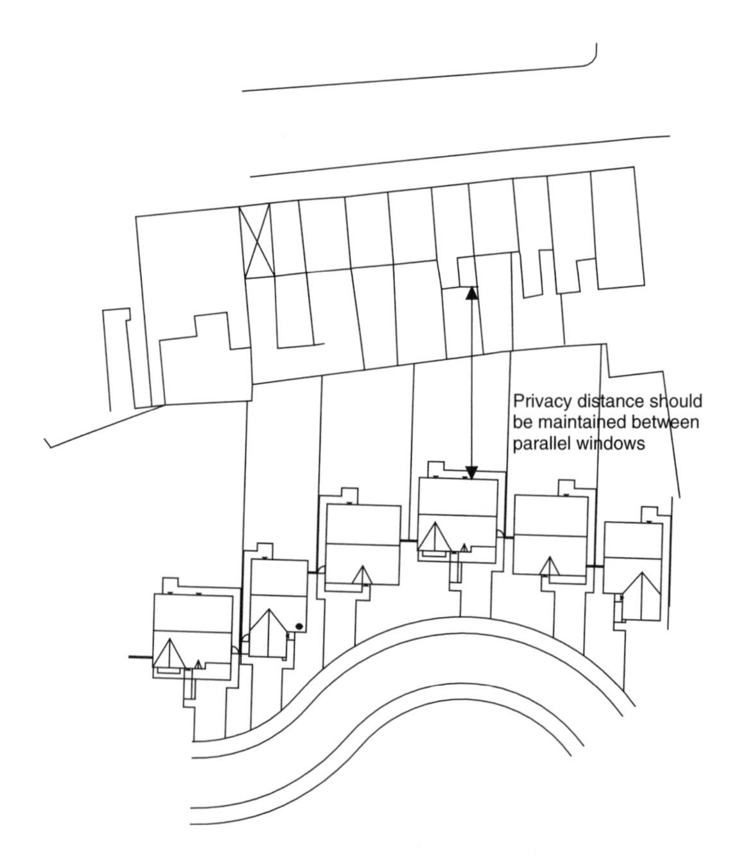

Fig. 8.12 Privacy distances.

BRIEF NOTES ON KITCHEN DESIGN

Generally

A well-planned kitchen undoubtedly increases the value of the property. I know because I sold my first house almost entirely upon the quality of the kitchen. To be honest the rest of the house didn't have a great deal to recommend it – except for the decoration.

However, unless I am requested to do so by my client, I do not provide internal kitchen layouts on my plans. The reason for this is because kitchen fitting has become a specialization in itself. However, a knowledge of the basics is essential. Figures 8.13 and 8.14 indicate a reasonably well-designed kitchen for a normal average sized home. However, don't forget, fashions change and new appliances appear every year.

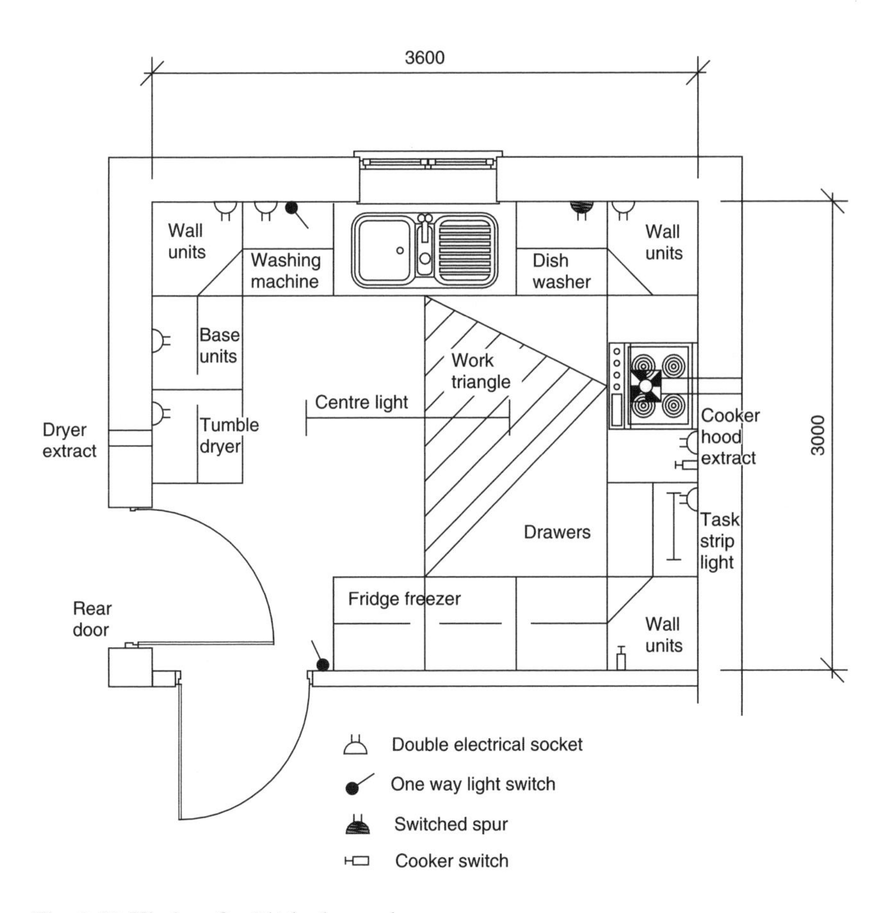

Fig. 8.13 Kitchen for 3/4 bedroom house.

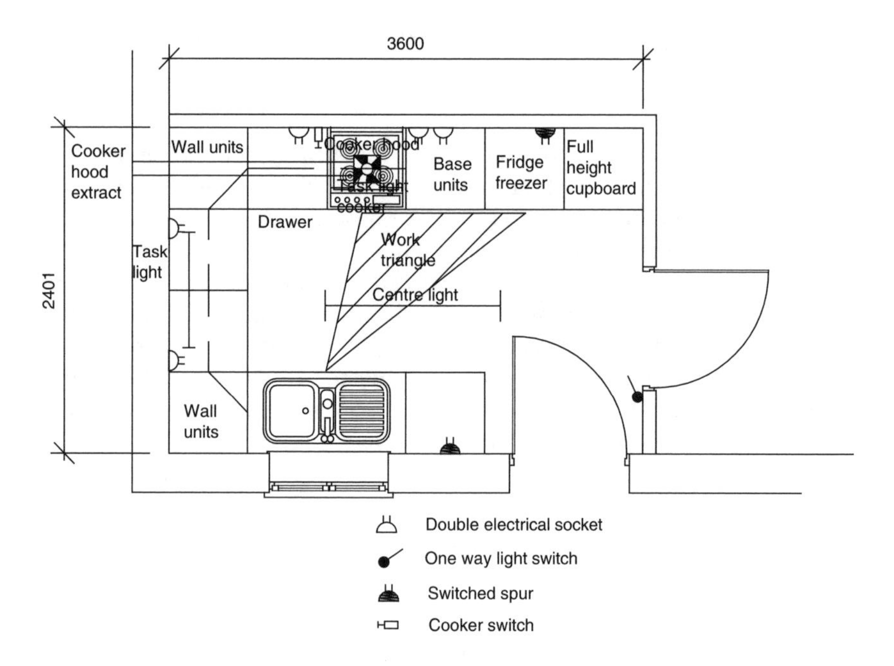

Fig. 8.14 Kitchen for 2/3 bedroom house.

Ideally, every kitchen should have space provided for a cooker/hob, refrigerator, dishwasher and freezer with hygienic (melamine faced or similar) worktops running around the kitchen linking the various work zones. Unless the house has a utility room, there should also be room in the kitchen to accommodate a washing machine and tumble dryer.

Design 'rules'

There are general 'rules' but it is not always possible to comply with them. They are:

(1) Work your layout on a 600 × 600 grid as this accommodates most appliances.
(2) A kitchen has three main activity centres

  (a) the food storage area (larder, fridge/freezer and/or store cupboard)
  (b) the sink
  (c) the cooker/hob
  (d) Whichever shape or size of your kitchen, try to make the kitchen as efficient as possible. Once you have decided where you intend to site the sink, preparation area and cooker, then draw three

lines to connect the three areas. This will usually result in a triangular shape (the work triangle). The smaller the work triangle, the shorter the walking distance will be between the three main working areas in your kitchen. If this walking distance looks too big, try to pull the three areas closer together.

(e) Rules of thumb for the basic work triangle are as follows

   (i) No one side of the triangle should be greater than 2.7 m (9 ft) or less than 1.2 m (4 ft).
   (ii) The triangle should not be interrupted by cabinets or a 'through route' for family members.
   (iii) The perimeter of the work triangle should measure no more than 8 m (26 ft) and no less than 3.6 m (12 ft).

(3) Place the fridge between the main entrance and the main cooking area so that family members can get easy access and shopping can be easily placed in the fridge. Placing the fridge next to the cooker should be avoided because the heat from the cooker could affect the temperature in the fridge (it's common sense).
(4) In most kitchens, the longest unit will consist of the sink, drainer and maybe a dishwasher, so it is best to decide where this is going first. Locate sinks under or near a window but no further away from a soil stack or gully than 2300 mm. With a single drainer unit, the drainer should be located on the side away from the cooker. This will allow quick access to the sink with hot pans without having to avoid the 'dryer up' or someone working at the drainer.
(5) If possible install a double bowl sink in new kitchens with one bowl capable of taking a waste disposal unit.
(6) The sink should not be placed in the corner of a room or up against a cupboard.
(7) Room should be left on the drainer side for someone to dry up.
(8) Cookers should not be placed under windows or wall cupboards.
(9) A cooker should be placed at least 300 mm away from a corner to allow easy access and door space.
(10) A cooker should not be placed in such a position that another person entering the room can open the door against the cooks back.
(11) If possible provide an extract or fan over a cooker.
(12) If possible, the cooker should be in line with the sink and not on the opposite side of the room.
(13) The cooker should never be separated from the sink by a door (either internal or external) because someone could open the door and walk while the cook was preparing hot food and there could be a serious accident or scalding.
(14) The washing machine should be near the sink. (If the property has a utility room a separate sink is usually provided.)

(15) Tumble driers should be on an outside wall and vented.
(16) Allow adequate space at hinge side of refrigerator/upright freezers because refrigerators/upright freezers have wide opening doors.

Electrical services

(17) Electrical services must comply with Approved Document P of the Building Regulations. Allow 30 amp control switch behind a free-standing cooker.
(18) Allow a 13 amp switch for each piece of major electrical equipment (e.g. dishwasher, washing machine, refrigerator, tumble dryer). These should be provided at working height and connected to outlets below the work surfaces. Wherever possible use twin socket outlets as the extra cost is minimal and this helps to prevent trailing flexes.
(19) If the funds are available and your client authorizes the expenditure allow artificial lighting in two forms:

   (a) Main background lighting.
   (b) Task lights to work surfaces in the main activity areas. (Task lights are usually hidden under wall cabinets.)

PART THREE
More on Building Control

9 More on building control

As explained in previous chapters, there are two local authority bodies that control simple development – the Planning Department and the Building Control Department. It is the latter that enforces the Building Regulations. (See below regarding Approved Inspectors.)

The Building Regulations are not uniform throughout the UK (Scotland and Northern Ireland have their own regulations). All references in this book are to the current Approved Documents that support the Building Regulations in force in England and Wales. It is essential that you obtain a copy of these Approved Documents if you intend to become deeply involved with plan preparation. Copies can be downloaded from the Internet. (Scottish or Northern Irish readers should obtain copies of their own regulations.)

Please also note, that until recently, Approved Document Part A – Structure used to contain timber tables (e.g. Floor joist spans/sizes etc.). This is no long the case. The tables now have to be obtained from TRADA or by using the Schedules provided in NHBC Standards. See later for details).

The important thing to remember is that the Building Regulations/Approved Documents seek to control how a building is constructed and not what it looks like.

THE APPROVED DOCUMENTS

The Approved Documents quite clearly state that they are 'intended to provide guidance' for some of the more common building situations. In other circumstances, alternative ways of demonstrating compliance with the requirements of the Building Regulations may be appropriate. There is no obligation to adopt any particular solution contained within the Approved Document. However, if you have not followed the guidance, it will be for you to demonstrate by other means that you have satisfied the requirements. In other words, the solutions indicated in the Approved Documents are not compulsory, but the onus of proving that you have complied with the Building Regulations falls on the applicant/agent.

When dealing with simple structures such as domestic extensions/new dwellings, following the laid down guidelines is usually the sensible approach unless for some reason alternatives have to be found.

The list of Approved Documents in support of the Building Regulations in force at the time of writing is as follows. (**Note**: Copies of the Approved Documents can be obtained from http:// www.planningportal.gov.uk.)

List of Approved Documents

- *Part A* Approved Document A – Structure (2004 edition)
  Because Approved Document A no longer contains any timber tables it is necessary to obtain a copy of 'Span tables for solid timber members in floor, ceilings and roofs for dwellings' from TRADA Technology. More details on TRADA Tables are given below.
  http:// www.trada.co.uk
  or
  NHBC Standards – The National Housebuilding Council www.nhbc.co.uk
- *Part B* Approved Document B (Fire safety) – Volume 1: Dwelling houses (2006 edition)
  Approved Document B (Fire safety) – Volume 2: Buildings other than dwellinghouses (2006 edition)
- *Part C* Approved Document C – Site preparation and resistance to contaminates and moisture (2004 edition)
- *Part D* Approved Document D – Toxic substances (1992 edition)
- *Part E* Approved Document E – Resistance to the passage of sound (2003 edition)
- *Part F* Approved Document F – Ventilation (2006 edition)
- *Part G* Approved document G – Hygiene (1992 edition)
- *Part H* Approved document H – drainage and waste disposal (2002 edition)
- *Part J* Approved document J – Combustion appliances and fuel storage systems (2002 edition)
- *Part K* Approved document K – Protection from falling collision and impact (1998 edition)
- *Part L – Dwellings* Approved Document L1A: Conservation of fuel and power (New dwellings) (2006 edition)
  Approved Document L1B: Conservation of fuel and power (Existing dwellings) (2006 edition)
- *Part L – Buildings other than dwellings* Approved Document L2A: Conservation of fuel and power (New buildings other than dwellings) (2006 edition)
  Approved Document L2B: Conservation of fuel and power (Existing buildings other than dwellings) (2006 edition)
- *Part M* Approved document M – Access to and Use of Buildings (2004 edition)
- *Part N* Approved document N – Glazing (1998 edition)

- *Part P* Approved document P – Electrical safety: Dwellings (2006 edition)
- *Regulation 7* Approved Document for Regulation 7 (1992 edition)

The effects of some of these Approved Documents will be considered in greater detail later in Part Four. However, as these documents are constantly being updated, it is essential for anyone involved in the building industry to keep abreast of the changes.

TRADA TABLES

As indicated above, the TRADA tables have now replaced the timber tables that were once included in Approved Document A. The tables provide timber sizing for the basic structural timber members in a conventional building. (e.g. Domestic floor joists, ceiling joists, rafters (banded into various slopes), purlins, and flat roof joists.)

The TRADA tables provide sizings for two timber grades C16 and C24. Tables also contain sketches which name the elements of structure and indicate how the tables should be used.

NHBC Standards

Although not strictly an Approved Document, NHBC Standards is worthy of a mention. The National Housebuilding Council regularily publish their own guide called NHBC Standards. This book (also on CD) is not a substitute for the Approved Documents. It is intended to supplement them and foster good building practice. I would recommend that you obtain a copy of NHBC Standards because it provides guidance, specifications, diagrams and solutions to many problems met with during a building operation. For instance, although it is not common to have to provide a fireplace and chimney in a small extension, occasionally it is. Chapter 6.80 provides excellent details of chimney construction.

Studying these details will also provide solutions to many problems met 'on site' and also increase your knowledge of basic building construction. NHBC Standards also provide extracts from 'Span tables for solid timber members in floor, ceilings and roofs for dwellings' from TRADA Technology mentioned above.

THE PURPOSE OF THE BUILDING REGULATIONS/APPROVED DOCUMENTS

When plans are sent into the Building Control Department at the local council, the Building Control Officer in charge of the application will check and ensure that the plans comply with the Regulations/Approved Documents. For example, the BCO will check to see that the plans indicate that the floor joists and roof joists are the correct size and that an adequate DPC has been shown built into brick-built external walls of habitable rooms. The BCO will also ensure that the submitted plans also indicate that the DPC is positioned correctly at minimum 150 mm (6 in) above external ground level. You can rest assured that the BCO will also work through a lengthy checklist to ensure that the plans are amended before any approvals are given. If the amendments are not made then the plans will be rejected.

As well as checking the plans, once work has commenced on site, it is usual for the Building Control Department to send a BCO out at regular intervals to ensure that the work complies with the approved plans, especially when the foundations are being constructed.

In order that the BCO knows when works are ready for inspection, there is a requirement that the builder/owner sends in stage notification cards at regular intervals. If these cards are not sent in and the work proceeds without the BCO being notified in due time, he or she can require the works to be opened up to ensure that the builder has complied with the Regulations. (**Note**: The BCO is not duty bound to make these inspections but the builder/owner has to let him or her know when the works are ready for inspection.)

Where appropriate, the BCO may also make observations or issue instructions to the builder should site conditions warrant it. (e.g. if firm or unfilled ground cannot be reached at normal depths, it is sometimes necessary for the builder to excavate the foundation trenches deeper than originally envisaged. Or, if the original assumptions shown on the approved plans need revising, the BCO may offer suggestions regarding the drainage system.)

In my experience, because they are very busy people, it is unusual for the Planning Officer to visit the site of a domestic extension during the construction period (they do normally make a site visit prior to granting approvals). However, most BCOs do liaise with Planning as the work proceeds. (This is not always the case. I have been informed by the NHBC that surveys that they have carried out indicate that only 50 per cent of local authorities have established arrangements for BCOs to check on planning conditions.)

However, if the authority concerned does have such arrangements, if the terms of the planning approval are contravened, you can be reasonably certain that the BCO will pass the information on to his colleagues in Planning and that someone from that department will swiftly appear and insist that the works be carried out in an approved manner. (**Note**: As both the Planning and Building Control Departments have recourse to law if their decisions are not obeyed, it is unwise to ignore all reasonable instructions.)

APPROVED INSPECTORS

Legislation contained within The Building Act 1984 made it possible for the supervision of building control matters to be carried out either by Local Authority Building Control Officers or by Approved Inspectors. The NHBC is one organization that is now carrying out the function of Approved Inspector.

I would also point out that the NHBC publish NHBC Standards from which I have quoted extensively and would recommend that you purchase these excellent publications which complement and amplify the Approved Documents/Building Regulations.

FULL PLANS/BUILDING NOTICE

The 1985 regulations brought in procedural changes. In the past, plans had to be submitted before work could commence on site. It is now possible to adopt one of two procedures which are as follows:

(a) Send a building notice to the local authority.
(b) Deposit full plans.

The idea behind the building notice system is that it will speed operations in as much that work can start on site virtually immediately without having to have plans approved prior to commencement. The problems that could reasonably be anticipated with option (a) the building notice – I would not recommend this procedure except exceptional circumstances. You as the designer/builder could be found liable in negligence if work carried out on the site does not comply with Regulations. The building notice system cannot be used for offices, shops and for designated buildings.

This book assumes that you will adopt option (b) and deposit full plans as this system protects the various parties against contravention of the Building Regulations. If you need to know more about the notice system, then it is usually possible to obtain the relevant information how it works from your local Building Control Department.

10 Is building control approval required?

Not every extension to a domestic property requires building control approval because like the Planning Department, Building Control do not wish to be flooded by hundreds of very minor applications that would be unlikely to create any major problems.

The types of exempt work most likely to be encountered when dealing with domestic alterations and improvements are as follows:

(a) porches
(b) conservatories (under 30 m^2 of floor area)
(c) covered ways open on at least two sides (under 30 m^2 of floor area)
(d) car ports open on at least two sides (under 30 m^2 of floor area).

Also, note the following:

(a) A conservatory or porch must satisfy the requirements of Part N of the Building Regulations. Safety glass must be incorporated where required by the Regulations.
(b) Just because the Building Regulations do not apply to these structures under the new regulations does not mean that Planning Permission is not required (refer to previous chapters). As with Planning, even if you believe that something that you are designing is exempt under Building Regulations, I would advise you to send a copy of the plan to the Building Control Department and ask for confirmation that building control approval is not applicable. Be safe, not sorry.

If your proposals do not come within the exempt category (kitchen extensions, lounge extensions and the like are not exempt), then you will need to apply for building control approval, and filling out the forms is dealt with in the next chapter.

11 Making the building control submission

ELECTRONIC SUBMISSIONS

As indicated in Chapter 6, the government and local authorities are now busily engaged in harnessing new technology. It is now possible to make a building control submission using Submit-a-Plan. As with the electronic planning applications, instead of printing out large number of drawings, the documents are sent in electronic form. Using this form of submission means that the applicant does not have to rely on the vagaries of 'snail mail'. In order to use the system you obviously need an internet connection. You then need to register on www.submitaplan.com

Submit-a-Plan is the LABC (Local Authority Building Control) National Portal for making applications to ANY local authority in England, Wales and Northern Ireland as well as reporting dangerous structures. The submission can be made direct to the intended local authority. It is also possible to track the progress of an (electronic) application.

So in brief, the key features of Submit-a-Plan are as follows:

(1) Other than the building control charge for the project, there are no hidden charges for using Submit-a-Plan.
(2) According to the Submit-a-Plan Website, the system works with all CAD applications and paper scans.
(3) Using the system eliminates sending multiple paper plans (when submitted electronically), saving both time, money and by eliminating paper it is environmentally friendly and helps to 'save the planet'.
(4) Because the system doesn't have 'office hours', it is possible to send an application at any time.
(5) For electronic submissions, it is also possible to track/progress applications online.

However, although electronic submission will very soon become standard practice, for the purposes of this book, I have used standard paper application forms in order to indicate what is involved.

FILLING IN THE FORMS

In order to obtain building control approval using the 'full plans submission' system, your plans have to be submitted to the relevant local authority together with a suitable application form. As with the planning application described before, under normal circumstances, local authorities have their own specially printed application forms. Each authority's forms vary slightly but the basic layout is the same.

I have reproduced a typical building control form and fee scales in Figs 11.1(a),(b) and 11.2(a)–(c). If you examine Fig. 11.1(a) you will note that a building control application form is very similar to a planning form and on the example shown requires answers to 11 questions. (**Note**: Although there are 12 questions on this form, questions 11 and 12 are either/or.) In this case I have signed the Full Plans Statement.

Question 1 asks for the name and address of the applicant. You should provide exactly the same information as you supplied on the planning form. The applicant is the person who wishes to have the works carried out (e.g. the homeowner).

Question 2 asks for the name of the agent. Once again, you should provide exactly the same information as you supplied on the planning form. If you are working for someone (you have prepared the plans for someone other than yourself) and are looking after the submission, then you are the agent and you must insert your name and address in order that all correspondence will be directed to you. If you do not give your name, any queries will be directed to the client. If you are not acting for someone (e.g. the application is for an extension to your own house), then there is no agent, so leave the box blank or cross it through. If you have a telephone number, provide it because if there are any queries, then someone can contact you easily and it helps to prevent delay.

Question 3 asks for the address applicable to the application. This address is not necessarily the same as the applicant's address. You or your client could be living at one address but own another property.

Question 4 asks for particulars of the development. Once again try to use the same description as you used on the planning application. Do not write down a massive rambling description.

Question 5 is aimed at establishing the use of the building. Remember, the Building Control Regulations vary from building type to building type. In this case, the extension is a simple home extension, so I have indicated that it is residential.

Questions 6 and 7 relate to building control fees. As with planning applications, fees are payable when an application is submitted. However, with Building Control the fees are usually paid in installments. Using the 'Full Plans' method, the plan submission fee is paid first. Once work commences, a second fee for on-site inspections then has to be paid.

(a)

THE BUILDING ACT 1984
THE BUILDING REGULATIONS 2000

Full Plans Submission or Building Notice

HALTON BOROUGH COUNCIL — HALTON LABC

Operational Director - Environmental and Regulatory Services, Environment Directorate, Halton Borough Council, Rutland House, Halton Lea, Runcorn, WA7 2GW.
E-mail: buildingcontrol@halton.gov.uk, www.halton.gov.uk

1 Applicant's Details (block capitals please)
Name MR & MRS J SOAP
Address 7 SOMEWHERE ROAD, HALTON H1 XYZ
Tel: ≈ ≈ Fax: ≈ ≈

2 Applicant's Agent (If applicable)
Name ANDREW R WILLIAMS
Address 437, WARRINGTON ROAD, RAINHILL L35 4LL
Tel: 0151 426 9660 Fax: andrew@ ------

3 Location of proposed work
7, SOMEWHERE ROAD, HALTON H1 XYZ

4 Brief description of proposed works or material change of use
SINGLE STOREY DINING & KITCHEN EXTENSION

5 Use of building (i.e. residential, office, shop, ect.)
Existing RESIDENTIAL Proposed RESIDENTIAL
Will the building be used for a use designated under Regulatory Reform (Fire Safety) 2005? YES ☐ NO ☑

Note: The section below must be completed, failure to do so may result in delay.

6 Type of work applied for (where applicable)
Number of dwellings No. N/A
Floor area of extension(s) sq. metres NE 40m²
Estimated cost of work £ N/A

7 Charges (See Guidance Note re: charges for information)
Plan charge £ 110-64 plus VAT £ 19-36 Total £ 130-00

8 Have you applied for planning permission? YES ☑ NO ☐
If yes please print Application No.

9 Conditional approval/minor amendments (where appropriate)
Do you consent to your plans being passed subject to conditions? YES ☑ NO ☐
Do you consent to minor amendments being made to the plan on your behalf? YES ☑ NO ☐

10 Extension of time (Full Plans submission)
Do you agree to extend the prescribed period for this application to 2 months? YES ☑ NO ☐

Note: Only one of the Statements below should be completed. See limitations of a Building Notice Use Guidance Notes over.

11 Statement Full Plans
This notice is submitted in accordance with Regulation 12(2)(b) and is given in relation to the work described above.
Signed: [signature] Date: ≈ ≈

12 Statement Building Notice
This notice is submitted in accordance with Regulation 12(2)(a) and is given in relation to the work described above.
Signed: Date:

Fig. 11.1(a),(b) Full plans submission form. (Reproduced with permission from Halton Borough Council.)

(b)

BUILDING REGULATION GUIDANCE NOTES

1 **What should be submitted to the local authority before commencing work**

Before you undertake any work, you, or your agent (i.e. builder, architect etc.) must advise the Local Authority either by submitting Full Plans for approval or through a Building Notice.

1. A Building Notice application may be submitted for most types of building work. However, a Building Notice may not be used:
- Where the proposed building work is intended to be put to a designated use for the purpose of the Regulatory Reform (Fire Safety) Order 2005, or
- Where the proposed work is within 3 metres of a public sewer, or
- A new building is erected fronting onto a private street

Currently all premises are designated, except single private dwellings.

2 **The Building Regulations Application**

Section 16 of the Building Act 1984 provides for the passing of plans subject to conditions.
The conditions may specify modifications to the deposited plans and/or that further plans shall be deposited.

2.1 Persons proposing to carry out building work, or make a material change of use of a building, are reminded that permission may be required under the Town and Country Planning Acts.

2.2 A completion certificate is available following satisfactory completion of the building work. The certificate must be requested, in writing, before the commencement of any works.

2.3 Persons carrying out building work must give written notice of commencement of the work at least 48 hours beforehand.

2.4 A Building Notice/Full Plans application shall cease to have effect from three years after the deposit date to the Local Authority unless the work has commenced before the expiry of that period.

3 **The Full Plans Procedure**

If you submit Full Plans the Local Authority will examine them and advise you of any necessary changes that are required to meet the Building Regulations. When considered satisfactory, a formal decision will be issued together with a copy of the approved plan and any relevant documentation and information on how to arrange for inspection of the work by a Building Control Consultant.

4 **The Building Notice Procedure**

Where the proposed domestic work includes the erection of a new building or extension this notice shall be accompanied by the following:

4.1 A block plan showing:-

1. *The size and position of the building, or the building as extended, and its relationship to adjoining boundaries, together with any other building within that curtilage;*

2. *The provision to be made for drainage of the building or extension;*

3. *Where it is proposed to erect the building or extension over a sewer or drain shown on the relative map of public sewers, details of the precautions to be taken to protect the sewer or drain.*

5 **Building Notice**

These notes are for general guidance only. Particulars regarding the submission of Building Notices are contained in Regulation 13 of the Building Regulations.

6 **Charges**

In respect of charges please see the Building (Local Authority Charges) Regulations and also the Halton scheme of charges.

Fig. 11.1 contd.

(a)

Schedule 2

Charges for certain small buildings, extensions and alterations.

TYPE OF WORK	PLAN CHARGE			INSPECTION CHARGE			BUILDING NOTICE		
	Charge £	V.A.T. £	Total £	Charge £	V.A.T. £	Total £	Charge £	V.A.T. £	Total £
1. Any extension of a dwelling (not falling within entry 4 & 5 below) the total floor area of which does not exceed 10 square metres, including means of access and work in connection with that extension.	110.64	19.36	**130.00**	113.19	19.81	**133.00**	223.83	39.17	**263.00**
2. Any extension of a dwelling (not falling within entry 4 & 5 below) the total floor area of which exceeds 10 square metres, but does not exceed 40 square metres, including means of access and work in connection with that extension.	110.64	19.36	**130.00**	217.02	37.98	**255.00**	327.66	57.34	**385.00**
3. Any extension of a dwelling (not falling within entry 4 & 5 below) the total floor area of which exceeds 40 square metres, but does not exceed 60 square metres, including means of access and work in connection with that extension.	110.64	19.36	**130.00**	327.66	57.34	**385.00**	438.30	76.70	**515.00**
4. Erection or extension of a detached or attached building which consists of a garage or carport or both having a floor area not exceeding 40 square metres in total and intended to be used in common with an existing building, and which is not an exempt building.	110.64	19.36	**130.00**	INCLUDED IN PLAN CHARGE			110.64	19.36	**130.00**
5. Erection or extension of a detached or attached building which consists of a garage or carport or both having a floor area exceeding 40 square metres but does not exceed 60 square metres in total and intended to be used in common with an existing building, and which is not an exempt building.	110.64	19.36	**130.00**	113.19	19.81	**133.00**	223.83	39.17	**263.00**

Minimum charges for loft conversions & extensions exceeding 60 square metres

Note : An estimated cost of work will be required

TYPE OF WORK	PLAN CHARGE			INSPECTION CHARGE			BUILDING NOTICE		
	Charge £	V.A.T. £	Total £	Charge £	V.A.T. £	Total £	Charge £	V.A.T. £	Total £
1. Any extension or alteration to a dwelling which consists of the provision of one or more rooms in a roof space.	110.64	19.36	**130.00**	217.02	37.98	**255.00**	327.66	57.34	**385.00**
2. Any extension of a dwelling, the total floor area of which exceeds 60 square metres.	Where an extension or alteration to a dwelling exceeds 60sqm, the plan charge and inspection charge or building notice charge must not be less than £438.30 (exc. VAT)								

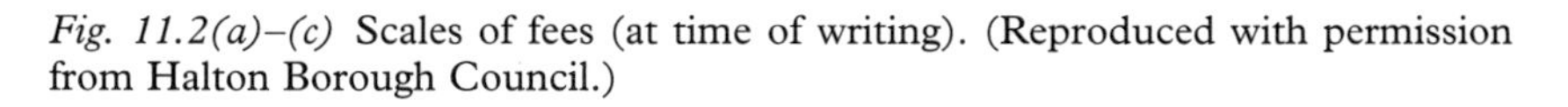

Fig. 11.2(a)–(c) Scales of fees (at time of writing). (Reproduced with permission from Halton Borough Council.)

(b)

Schedule 3

Charges for work based on estimated cost of other work not covered in schedule 1& 2.

TOTAL COST	FULL PLANS						BUILDING NOTICE		
	PLAN CHARGES			INSPECTION CHARGES			BUILDING NOTICE CHARGES		
£	Charge £	V.A.T. £	Total £	Charge £	V.A.T. £	Total £	Charge £	V.A.T. £	Total £
0-1000	50.00	8.75	**58.75**	Not applicable			50.00	8.75	**58.75**
1001-2000	100.00	17.50	**117.50**	Not applicable			100.00	17.50	**117.50**
2001-5000	165.00	28.88	**193.88**	Not applicable			165.00	28.88	**193.88**
5001-6000	43.50	7.61	**51.11**	130.50	22.84	**153.34**	174.00	30.45	**204.45**
6001-7000	45.75	8.01	**53.76**	137.25	24.02	**161.27**	183.00	32.03	**215.03**
7001-8000	48.00	8.40	**56.40**	144.00	25.20	**169.20**	192.00	33.60	**225.60**
8001-9000	50.25	8.79	**59.04**	150.75	26.38	**177.13**	201.00	35.18	**236.18**
9001-10,000	52.50	9.19	**61.69**	157.50	27.56	**185.06**	210.00	36.75	**246.75**
10,000-11,000	54.75	9.58	**64.33**	164.25	28.74	**192.99**	219.00	38.33	**257.33**
11,001-12,000	57.00	9.98	**66.98**	171.00	29.92	**200.92**	228.00	39.90	**267.90**
12,000-13,000	59.25	10.37	**69.62**	177.75	31.11	**208.86**	237.00	41.48	**278.48**
13,000-14,000	61.50	10.67	**72.26**	184.50	32.29	**216.79**	246.00	43.05	**289.05**
14,000-15,000	63.75	11.16	**74.91**	191.25	33.47	**224.72**	255.00	44.63	**299.63**
15,001-16,000	66.00	11.55	**77.55**	198.00	34.65	**232.65**	264.00	46.20	**310.20**
16,001-17,000	68.25	11.94	**80.19**	204.75	35.83	**240.58**	273.00	47.78	**320.78**
17,001-18,000	70.50	12.34	**82.84**	211.50	37.01	**248.51**	282.00	49.35	**331.35**
18,001-19,000	72.75	12.73	**85.48**	218.25	38.19	**256.44**	291.00	50.93	**341.93**
19,001-20,000	75.00	13.13	**88.13**	225.00	39.38	**264.38**	300.00	52.50	**352.50**
20,000-100,000	£75.00 Plus £2.00 excluding V.A.T for each £1000(or part thereof)			£225.00 Plus £6.60 excluding V.A.T for each £1000(or part thereof)			£300 Plus £8.00 excluding V.A.T for each £1000(or part thereof)		
Over 100,000	Contact Halton Building Control Consultancy								

Charges for replacement windows

	Charge £	V.A.T. £	Total £
Under £5000	£51.06	£8.94	**£60.00**
Work over £5000 will be charged at 35% of the Building Notice Charge plus VAT			

Fig. 11.2 contd.

(c)

Replacement Windows

All replacement glazing now falls within the scope of the Building Regulations.

Anyone who installs replacement windows or doors will need to comply with thermal performance standards. One of the main reasons for this change is the need to reduce energy loss. The Building Regulations have controlled glazing in new buildings for many years but they represent only a very small percentage of our total building stock. It is also essential to improve the performance of the much larger numbers of existing buildings if we are to meet increasingly stringent National and Global energy saving targets.
When the time comes to sell your property, your purchaser's surveyors will ask for evidence that any replacement glazing installed after 1st April 2002 complies with the new Building Regulations. So you will need a certificate from the Local Authority saying that the installation has approval under the Building Regulations.

As an alternative some contractors will be able to offer self-certification under the FENSA scheme.

The FENSA Scheme (domestic work only)

The scheme allows installation companies that meet certain criteria to self-certify that their work complies with the Building Regulations. The scheme is known as FENSA, which stands for Fenestration Self-Assessment. It was set up by the Glass & Glazing Federation, in association with all key stakeholders, and meets with Central Government approval. A sample of the work of every installer will be inspected by FENSA appointed inspectors to ensure standards are maintained. FENSA issue certificates to householders confirming compliance. This scheme does not apply where the work is of a structural nature such as replacing bay windows, lintels or enlarging openings.
Any installation done by a firm which is not registered to self-certify, is of a structural nature or done as a DIY project by a householder, will need full Local Authority approval under the Building Regulations. Local Authorities will know of all the approved installers in their area and will be able to identify unauthorised work very easily.

You should note that you, as the house owner, are ultimately responsible for ensuring the work complies with the Building Regulations.

Before you sign a contract to buy replacement glazing, be sure to ask whether the installer is able to self-certify and decide if you want them to do so. If not, either they, or you, will need to make an application to Halton Borough Council's Building Control Consultancy for approval under the Building Regulations and pay any relevant charges.

Halton Borough Council,
Environmental Health and Planning Department,
Environment Directorate,
Rutland House, Halton Lea, Runcorn,
WA7 2GW, Telephone: 0151 424 2061,
E-mail: building.control@halton.gov.uk,
www.halton.gov.uk

HALTON
LABC

Halton Building Control Consultancy/charges/may2004

Fig. 11.2 contd.

Figure 11.2 indicates the fee scales applicable to building control applications as at the time of writing. Unlike planning application fees, VAT is charged on building control fees.

(**Note**: If you are carrying out work for a disabled person, fees are sometimes waived, but proof has to be provided to substantiate the disablement (e.g. a doctor's letter). Work that is obviously not entirely for the use of a disabled person will attract a fee.)

Question 8 asks if you have applied for planning permission. It is designed to ensure that planning procedures are not forgotten.

Question 9 Conditional approval/minor amendments. By ticking these boxes, it allows the BCO checking your plans to correct simple errors on the plans or grant conditional approval for proprietary items such as roof trusses.

The reason for conditional approval is that specialist manufacturers will not provide calculations until they are given an official order from the builder. Obviously, a designer is not in a position to give an official order. Rather than cause delay, the BCO grants a conditional approval on the understanding that the outstanding information is provided by the builder in good time. When dealing with question 9, I usually answer 'Yes'. If you prevent the BCO from having the right to condition the approval/make minor amendments, the officer might be forced to reject the plans because you have not allowed him or her to do otherwise.

Question 10 By signing this section, you agree to give the local authority additional time to consider the application. I always mark this 'Yes'. If you refuse to give the BCO additional time and the plans are not quite right then he or she will be forced by law to reject the application if amendments are needed and the amendments are not made quickly enough.

Question 11 As explained before, I normally advocate the full plans system.

Question 12 This procedure does not involve the passing or rejecting of plans. It enables building work to get underway quickly. But there are risks involved with the system. Using the notice system, the builder and the property owner are 'building at their own risk'. Someone who starts work without 'full approval' and without approved plans to guide them might carry out work incorrectly and be forced to rectify the defects at their own cost.

SENDING THE DOCUMENTS IN

When submitting your documentation read the forms carefully. Irrespective *of what* may be laid down by the Regulations, most Authorities have their own requirements regarding numbers of forms and plans *that they* require. The forms issued by your Building Control Department usually stipulate the numbers of copies *of each* document that they need. If you do not provide the correct number, you could delay your client's application.

Do not forget to send a cheque for the relevant plan fee. The authority will not accept the application unless the fees are included.

QUERIES RAISED BY BUILDING CONTROL

Unlike the Planning Department, who rarely contact the applicant/agent once the plans are lodged (unless there is a serious non-conformity with established policy) it is virtually certain that you will receive a notification from the Building Control Department informing you that your plans are defective and do not comply with the Building Regulations.

When you receive a notification of this nature, there is a tendency for the inexperienced agent or householder to be either insulted or very concerned. As I said before, queries are the norm because the Building Regulations are very complex, and even someone who has a very good working knowledge of the Building Regulations can omit to include vital information. If a plan passes without any queries then you can count yourself as being very lucky.

Sometimes if amendments are only minor a letter may be issued requesting authorization to amend the plans.

ACTION TO BE TAKEN UPON RECEIPT OF A NOTICE OF NON-COMPLIANCE

I would advise that you take the following action:

(1) Telephone the BCO and clarify any points which are not obvious. Most forms tend to be couched with vague references to the Building Regulation clauses. Find out what is not considered correct and try to rectify the omission. Sometimes your plan may actually cover the point in question and the officer has missed it because he or she has not had the time to fully check the application (everyone is human and errors can occur).

(2) Fill in the extension of time form and post it back (if you have not already granted an extension on the form).

(3) Amend the master copy of your plans immediately and send revised sets of copy plans to the BCO with a covering letter stating clearly the application number that has been issued to your submission.

RECEIVING A REJECTION NOTICE

There will be occasions when your plans will just be stamped rejected and sent back. When this happens, once again, there is a strong temptation for the inexperienced homeowner/agent to dress up in sackcloth and ashes and bewail his or her lot, especially if no notice informing you of non-compliance has been received.

Although BCOs very rarely admit to being overloaded with work, they are sometimes forced to 'defend their corner' by clearing out those applications which either cannot be dealt within the time available or which require a large number of amendments making.

A rejection notice does not automatically mean that the BCO considered the plan that you produced was so terrible that it was beyond redemption.

It should be borne in mind that Building Control are obliged to pass or reject an application within a set timescale (five weeks, or eight weeks if an extension of time has been granted).

ACTION TO BE TAKEN IF PLANS ARE REJECTED BY BUILDING CONTROL

I would advise that you take the following action:

(1) Once again, telephone the BCO and clarify any points which are not obvious. (Do not be rude or offensive. The person on the other end of the phone is only doing his or her job.) Find out what is considered incorrect and try to rectify the omission/mistake.
(2) Resubmit the application as quickly as possible after making any amendments, but make sure that you indicate that the application is a resubmission.

Resubmissions do not normally require a fee to be sent with them unless the plans submitted are nothing like the original application.

PART FOUR
Building Construction

12 General information

BUILDING CONSTRUCTION TERMINOLOGY

Like any other industry, the building industry has developed its own peculiar terminology for components that form a finished structure. These terms have grown up over the centuries and are the ones used in technical documentation, including the Building Regulations. If you refer to Figs 12.1 and 12.2 some of the major elements are indicated. Some of my readership may be familiar with most of the terms but obviously, if you are not, then you should examine the details in order that you can understand the basics.

As you will note, Fig. 12.1 shows a typical house with the parts of the elements 'peeled back' in order to view the underlying detail. Figure 12.2

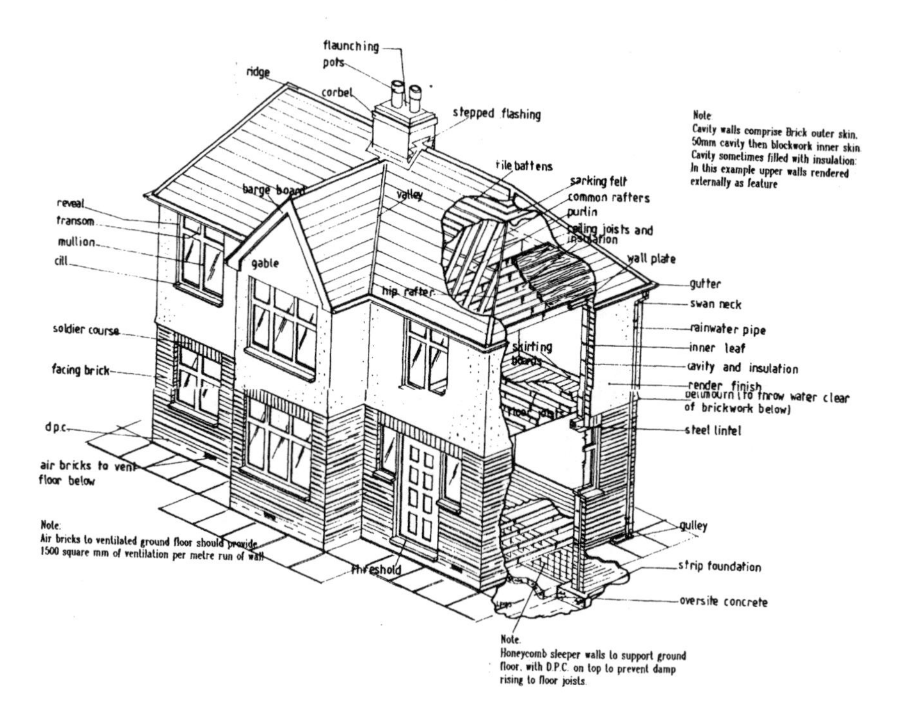

Fig. 12.1 'Peeled back' detail of typical house.

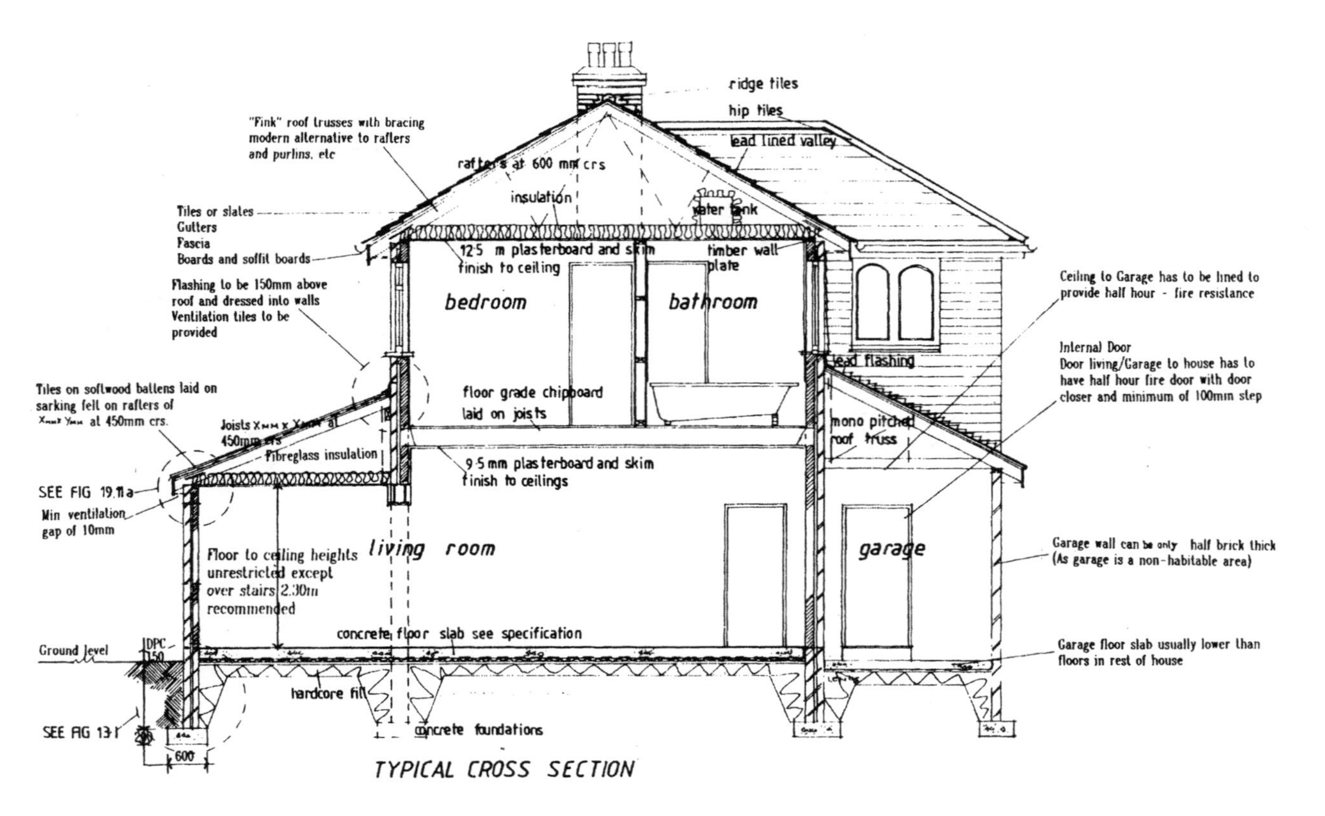

Fig. 12.2 Typical cross section.

is a cross section of a typical building. More detailed sketches have been provided later in the book when dealing with major elements (e.g. foundations, roofs etc.). When preparing a plan, it is essential that you use standard building terminology and are not tempted to invent your own. I have seen plans prepared by people who are not totally conversant with building construction describing say a purlin, as say a roof beam. This can be confusing to builders and BCOs. The use of non-standard terminology will also serve to highlight your inexperience.

For those unfamiliar with any building terminology, I have listed some of the major elements in a building and provided an outline description in Appendix C.

However, the list is not exhaustive and I would recommend that you buy a Building Dictionary.

NOTES CONCERNING NEW APPROVED DOCUMENT B VOLUME 1 OF THE BUILDING REGULATIONS – FIRE SAFETY (2006 EDITION)

Part B Volume 1 deals with Dwelling Houses (not flats) and covers a variety of different types and sizes of dwelling. As the terms dwelling/dwelling house seem rather archaic, I shall just use the term house(s) from now on. With the exception of loft conversions, *I have deliberately concentrated on the simple domestic extension to houses that do not have upper floors above 4.5 m above ground level.*

The notes provided are only a general guide and reference must be made to Approved Document B on larger projects.

Fire resistance

Fire is one of the major destroyers of buildings. The Great Fire of London and the Summerland disaster testify to this. In order to mitigate the effects of fire, precautions have to be taken. A standard two-storey dwelling house has to have a minimum fire resistance of 30 minutes. As plasterwork is usually specified on walls and ceilings, this usually provides for the requirement. In order to ensure that items such as steel beams comply with the 30-minute rule, they usually have to be encased (usually in plasterboard) to prevent them distorting for the first 30 minutes of the fire. The reasoning for this is that it enables the occupant of the house a full 30 minutes to escape/to be rescued before the fire begins to destroy major structural elements of the house and the inevitable collapse beings.

Windows/doors near boundaries

If you are building an extension with normal brick and block walls, because brickwork are unlikely to burn in the event of a fire, you can normally build up to a boundary.

BUT WHAT IF THE WALL HAS DOORS AND WINDOWS IN IT?

If you introduce doors and windows into a wall facing a neighbouring boundary, special rules apply. See table below:

Minimum distance to boundary	Maximum unprotected area (area of windows and doors)
Under 1 m (1000 mm)	1 m^2 maximum
1 metre (1000 mm)	5.6 m^2
2	12 m^2 #
3	18 m^2 #
4	24 m^2 #
5	30 m^2 #
6	No limit #

(**Note**: # It is unlikely that you will ever use such large areas in a home extension because thermal regulations (Approved Document Part L) restricts glazed areas to prevent heat loss and global warming.)

Roof lights

It is very common for modern extensions to incorporate unwired roof lights (e.g. Velux). These do not normally cause a problem from the fire point of view. Thermoplastic roof lights cannot exceed 0.5 m^2 and have to be spaced at 3 m minimum centres.

Smoke alarms

Modern houses are built with smoke alarms installed. Because the Building Regulations are not retrospective, there is no requirement for an existing house to be brought into line. However, if an extension is built containing a habitable room then the Building Regulations use the extension as a means of bringing the subject property into line with new houses and smoke/heat alarms have to be installed.

The smoke/heat alarms have to comply with the following:

(1) Be installed to BS 5839–6:2004. All smoke alarms should have a standby power supply.

(2) They should normally be positioned in corridors between bedrooms and kitchens and living rooms (being the places where fires are most likely to start). There has to be a *smoke alarm* in circulation spaces (corridors) within 7.50 m of the doors of every habitable room and 3 m from bedroom doors.
(3) There has to be at least one alarm per floor.
(4) Individual smoke/heat alarms have to be interlinked so that if one goes off, all units sound the alarm.
(5) Smoke alarms should not be fitted in very hot areas where they could give false alarms.
(6) Where a kitchen is not separated from the stairway/circulation space by a door, there should be a heat alarm/heat detector in addition to the other smoke alarms.

Compartmentation of attached garages

Where a domestic garage is connected to a dwelling, the garage is compartmentalized (isolated from the rest of the house) by having a minimum 30-minute firewall between the house and the garage (a normal cavity wall exceeds 30 minutes fire resistance).

If there is a door connecting the house to the garage, the door must be an FD30S. (The reference indicates that the door has a 30-minute fire rating and that it has smoke seals.) Such a door also requires a self-closing device fitted (e.g. Perko closer or Briton door closer).

The door must also be 100 mm above the floor to prevent fuel entering the house in the event of a spillage/fire. Alternatively, the garage floor can be laid to falls. (The falls being away from the fire door.) **Note**: The Building Regulations quote the alternative the other way around. However, I personaly prefer a 100 mm step.

Means of escape

For simple domestic extension to houses that do not have upper floors above 4.5 m above ground level, the most important provisions are the following:

Ground floor

With the exception of kitchens, all habitable rooms on the ground floor shall:

(1) Open onto a hall leading the front door (or other suitable exit), or,
(2) A window or external door which has an unobstructed operable area of at least 0.33 m^2 (and 450 mm in length and breadth), not more than 1.10 m above floor level and leading to a safe area well away from the fire.

First floor

With the exception of kitchens, all habitable rooms on upper storeys only served by one staircase shall

(1) Be provided with a window or external door which has an unobstructed operable area of at least 0.33 m^2 (and 450 mm in length and breadth), not more than 1.10 m above floor level and leading to a safe area well away from the fire. On the window, the cill height should not be less than 800 mm, fitted with child-resistant release catches and designed to remain in the open position without having to be held open by the person attempting to escape. Or,
(2) Have direct access to a protected staircase (surrounded by walls and doors providing half hour fire protection).

Galleries

A gallery is a raised area or platform around the sides or back of a room. See Figs 12.3 and 12.4.

To comply with Approved Document B1, a gallery should be provided with

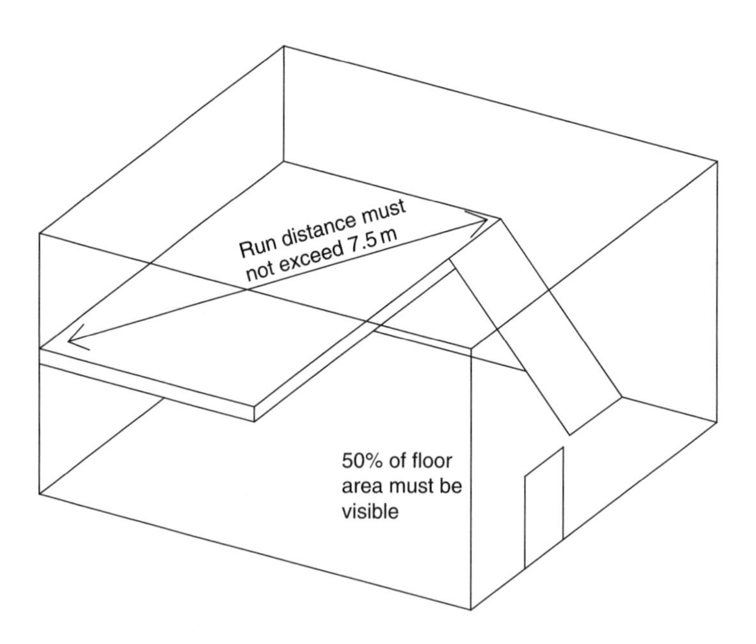

Fig. 12.3 Gallery floors with no escape window.

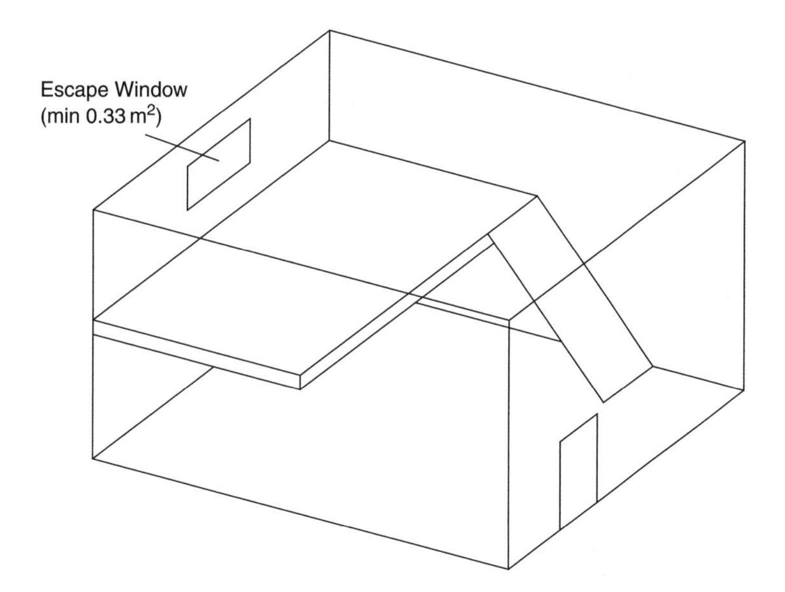

Fig. 12.4 Gallery floors with escape window.

(1) an alternative exit (a door leading off the gallery)
(2) where the gallery floor is not more than 4.5 m above ground level an approved escape window can be used (see Fig. 12.4 for details).

Alternatively

If the above requirements are not complied with (see Fig. 12.3), the gallery should:

(1) Overlook at least 50 per cent of the floor area below (presumably so that whoever is on the gallery floor can see if a fire has broken out).
(2) The foot of the stair to the gallery shall only be 3 m from the exist door of the room (to ensure easy exit in the event of a fire).
(3) Any point on the gallery and the head of the access stair shall not exceed 7.5 m.
(4) Any cooking facilities in the room containing the gallery should either be enclosed or positioned in such a position not to prejudice escape down the stairs.

Loft conversions

The 2006 edition of Approved Document B – Volume 1 alters the way that loft conversions were treated. Prior to the new regulations, roof windows/gable windows could be used for escape from loft conversions. Although this appears to still be possible if one is converting a bungalow roof, in a loft conversion on a two-storey house this is no longer an option. The only means of escape now allowed for three-storey houses is via a protected stair (surrounded by walls and doors providing half hour fire protection).

Inner rooms

An inner room is a room whose only escape route is through another room. The main risk with the inner room situation is that if a fire starts in the adjacent room (access room), the person inside is trapped. The only acceptable arrangements for inner rooms are where the inner room is

(1) a kitchen
(2) a utility/laundry room
(3) a dressing room
(4) a bathroom/WC/shower room
(5) a room not more than 4.5 m above ground level which has an escape window described above
(6) a gallery that complies with the previous section.

NOTES CONCERNING NEW APPROVED DOCUMENT L1B – CONSERVATION OF FUEL AND POWER (EXISTING BUILDINGS, 2006 EDITION)

Preamble

A few decades ago, the possibility of nuclear destruction was the secret fear. The current political obsession is global warming and all over the world, stable doors are slamming shut. Unfortunately, the horse bolted years ago but no one noticed. In order to control the fleeing nag, Approved Document L has been upgrade yet again. And its now a lot bigger and it has been divided up into four sections. *L1B deals with existing dwellings (and extensions)*. As the terms dwelling/dwelling house seem rather archaic, I shall just use the term house(s) from now on.

U-values

In Approved Document L1B, guidance is given concerning the new thermal requirements of an extension to a house. These thermal requirements are shown as ***U*-values**.

Defining the term *U*-value

A *U*-value is a measure of heat loss. It is expressed in W/m^2 K, and shows the amount of heat lost in watts (W) per square metre of material (e.g. wall, roof, glazing etc.), when the temperature (K) is one degree lower outside.

The lower the *U*-value, the better the insulation provided by the material.

Useful Weblinks

http://www.knaufinsulation.co.uk
http://www.celotex.co.uk/u_**value**.php

The requirements of approved document L1B

A word of warning!

If and when you open a copy of L1B, don't go straight for Table 1!

For reasons best known to themselves, the authors of LIB seemed to have strived to turn their readers into contortionists. If you have obtained a copy of the Approved Documents, the first table to refer to is Table 4 on page 22 which indicates the basic thermal requirements for a simply designed extension and Table 2 which provides information on windows and doors. There are three ways to comply with the regulations:

OPTION 1 (USING THERMAL ELEMENTS THAT MEET THE STANDARDS)

Using Option 1, the elements of structure have to comply with the *U*-values given in Tables 2 and 4 of L1B:

New Extension (using Option 1)

Information below is an amalgam of Tables 2 and 4 of L1B

Element of structure	*U*-value (W/m^2 K)
Wall	0.30
Windows, roof windows and rooflights	1.8 (or centre pane value 1.2)
External doors with more than 50 per cent glazing	2.2 (or centre pane value 1.2)
Other external doors	3.0
Pitched roof – Insulation at ceiling level	0.16
Piched roof – Insulation at rafter level	0.20
Flat roof or roof with integral insulation (e.g. SIPs construction – see later)	0.20
Heat loss floors	0.22

- If Option 1 is adopted, the area of the doors and windows combined should not exceed 25 per cent of the floor area of the extension + the area of any existing doors and windows that have been enclosed by the new extension.
- Windows, roof windows, rooflights and doors are controlled fittings and have to be draft proofed.
- Where heating and hot water are provided, these controlled services have to comply with clause 35 to 38 of L1B.
- In addition to the above, thermal bridges and unwanted air leakage have to be avoided by using robust details.

EXAMPLE OF DOOR/WINDOW CALCULATION

If you have an extension with an internal floor area of $10\,m^2$ and the existing doors and windows of the original house (and now enclosed within the extension) were $4\,m^2$, the permissible door/window area for Option 1 solution would be $2.5\,m^2$ ($25\%^{\text{of floor area}} + 4\,m^{2\text{(of original windows within extension)}} = 6.5\,m^2$).

Replacement Elements (Using Option 1)

If during the building works some surrounding thermal elements of the existing house are replaced then the following must be observed. Information below is an amalgam of Tables 2 and 4 of L1B

Element of structure	*U*-value (W/m² K)
Wall	0.35
Windows, roof windows and rooflights	2.0 (or centre pane value 1.2)
External doors with more than 50 per cent glazing	2.2 (or centre pane value 1.2)
Other external doors	3.0
Pitched roof – Insulation at ceiling level	0.16
Piched roof – Insulation at rafter level	0.20
Flat roof or roof with integral insulation (e.g. SIPs construction – see later)	0.25
Heat loss floors	0.25

- Windows, roof windows, rooflights and doors are controlled fittings and have to be draft proofed.
- Where heating and hot water are provided, these controlled services have to comply with clause 35 to 38 of L1B.
- In addition to the above, thermal bridges and unwanted air leakage have to be avoided by using robust details.

OPTION 2 (USING THERMAL ELEMENTS THAT MEET THE STANDARDS)

First, let's start with a question.

Why is Option 2 needed?

The reasoning behind Option 2 is that Option 1 is very inflexible. What if a client wasn't happy with 6.5 m^2 calculated above? Say he wanted 7 m^2 of doors and windows in his extension.

The way to solve the problem is by using Option 2. Option 2 compares the proposed extension with the extension designed under Option 1 and if the heat loss is no greater then it is acceptable.

Example

If you have an extension with an internal floor area of 10 m^2 and the existing doors and windows of the original house (and now enclosed within the extension) were 4 m^2 the permissible door/window area for Option 1 solution would be 2.5 m^2 (25%$^{\text{of floor area}}$+ 4 $m^{2(\text{of original windows within extension})}$ = 6.5).

But the client wants 7 m^2 of doors and windows.

Heat loss to original Option 1 extension	Area	*U*-value	Total	
Wall	22.50	0.3	6.75	
Windows	4.50	1.8	8.10	
External doors (>50% glazed)	2.00	2.2	4.40	
Pitched roof	10.00	0.16	1.60	
Floor	10.00	0.22	2.20	
Total Heat Loss			23.05	
Heat loss for Option 2 extension (Heat loss to original Option 1 extension)	Area	*U*-value	Total	Difference
Wall (wall insulation increased)	22.00	0.25	5.50	−1.25
Windows	5.00	1.8	9.00	+0.90
External doors (>50% glazed)	2.00	2.2	4.40	0.00
Pitched roof (roof insulation increased)	10.00	0.14	1.40	−0.20
Floor (floor insulation increased)	10.00	0.2	2.00	−0.20
Total Heat Loss			22.30	−0.75

If you compare the two sets of calculations by super insulating some elements of structure, it has been possible to comply with Building Regulations whist at the same time giving the client larger windows. In fact, in this example, the insulation standard offered in Option 2 is better than Option 1.

Note: The *U*-values used for this exercise have to comply with the requirements of Table 1.

OPTION 3 (COMPARING CO^2 EMISSIONS)

Option 3 is similar to Option 2 except that SAP 2005 (The government's 'Standard Assessment Procedure' for Energy Rating Buildings) has to be used to calculate the CO^2 emissions from the above extensions and compare the two. As SAP rating is a specialism in itself, it is beyond the scope of this book.

CONSERVATORIES, SUBSTANTIALLY GLAZED EXTENSIONS AND THE TV KITCHEN

General comments on conservatories

Although there have always been companies like Amdega around to construct high quality conservatories for the well heeled, it wasn't so long ago that the conservatory for most people was nothing more than a tacked on, home-made lean-to. Either that, or it was a single glazed aluminium structure. For the most part, the older style conservatory that most people had was basically nothing more than an attached greenhouse.

And that in a nutshell is how most BCOs (and the Building Regulations) view conservatories. They are treated as an add-on and not considered as a part of the main house.

There is, however, an advantage to this attitude. As conservatories are not considered part of the habitable structure of the house, as long as the floor area does not exceed 30 m^2, the Building Control Department are not that interested in them. (There is a requirement that the glazing in conservatories comply with Approved Document N — Glazing. Also note below the comments on thermal separation below.)

The definition of a conservatory

Clause 22 of Approved Document L1B deals with conservatories and substantially glazed extensions. As indicated above, conservatories under 30 m^2 (which is larger than most conservatories on a small to medium domestic house) do not require building control approval.

However, note the definition of a conservatory below:

A conservatory is defined as an extension to a building which has not less than three quarters of the roof area and not less than half of its external wall area made from translucent material and is thermally separated from the house. (Usually a set of external quality French doors or Patio doors are used as separators between the house and conservatory.)

The last proviso is important. If there is no adequate thermal separation, then because heat is escaping from the house, a non-separated conservatory has to comply with the full effect of Approved Docuement L1B. (i.e. The walls and roof are treated like windows and it is necessary to carry out an Option 2 calculation (or Option 3) as described above.)

Substantially glazed extensions (sunrooms)

A sunroom is basically a conservatory without a glazed roof. Until recently, sunrooms were given a hard time by Building Control because they were not recognized as being different from any other habitable home extension.

However, times appear to have changed. As indicated above, clause 22 of Approved Document L1B recognizes with substantially glazed extensions and directly links them to conservatories. If I am reading the new clause correctly, as long as the sunroom can be proven to be as thermally efficient as a conservatory of the same size, then it can be treated like a conservatory.

A WORD OF WARNING!

To gain the concession of being treated like a conservatory, a sunroom has to be thermally separate just like a conservatory. If they are not kept thermally separate then the sunroom will be treated exactly like any other form of home extension.

The TV kitchen conservatory

Unfortunately, many homeowners want what they see on TV.

When Dallas was on TV, there was a certain type of homeowner that wanted unnecessary steps inserting into floor plans 'just for flash'. They seemed to forget that when Granny called, she was virtually guaranteed to trip over the unexpected hazard, go arse over tip and wind up in hospital.

The TV kitchen conservatory is a typical example of seeing and wanting.

As it is not uncommon to see some famous TV personalities preparing an elaborate meal in what appears to be a conservatory, many homeowners want to place their kitchens partially inside the conservatory. But beware – I have never quite worked out how the TV personality manages to avoid severe condensation problems.

As indicated above, from the Building Control point of view, if there is no thermal separation between the main room and the conservatory, the conservatory will be/is treated like a giant window for Building Control purposes and it will be necessary to use Option 2 (or Option 3) described above to prove that the house complies with the requirements of modern thermal efficiency. If the conservatory is a large one, this will not be an easy exercise.

13 The foundations

GENERALLY

The foundations of the house are probably one of the most important parts of the property because without an adequate base the property will quickly become unstable. But the foundations are only as strong as the substrata of ground upon which they are built. Therefore, the substrata underneath the foundations should comply with the following:

(a) Be strong enough to sustain the loads put on it by the building. (If it is not, then a different type of foundation or wider foundation is needed to spread the load.)
(b) Not contain sulphates or other deleterious matter likely to destroy the concrete in the foundations. (Alternatively, the concrete in the foundations must be capable of resisting the effects of the sulphates or other deleterious matter.)
(c) Not be susceptible to the effects of frost action. (Clay can expand if affected by frost. That is why gardeners like to ridge up clay soil in cold weather so it breaks up. If the subsoil is of a type susceptible to frost action then the underside of the concrete foundations must be built sufficiently deep that the soil beneath them will not freeze.)

However, foundation design can present a problem for anyone designing small domestic extensions because, unlike larger projects, you, the designer, are very much on your own. There are no other consultants such as engineers on hand to make soil tests and to advise you about ground conditions in the area.

On page 36 of Approved Document A (of the Building Regulations), the subsoils upon which your foundations are likely to be built are classified into seven types (e.g. rock, clay, sand) and there is also a list of field tests that you can carry out. For instance, silt is defined as 'fairly easily molded in the fingers and readily excavated'.

Well, that's great, except that even a lay person knows that foundations are built below ground level. So, what is the point in attempting to assess the subsoil condition from ground level?

You could of course visit your client armed with a pickaxe and spade and start by digging trial holes in their garden (trial holes are pits about

1meter square excavated as deep as necessary to hit firm ground, and are usually excavated on larger sites in order to test soil conditions), but in my experience, most homeowners do not want their builder/surveyor digging up their gardens months in advance of work starting on site, especially as they know that they will have enough disturbance once work actually starts.

In any case, even if you dig trial holes, it does not necessarily mean that they will give the full picture. It is not unknown for the original builders to construct houses on sites that are far from perfect. Whilst doing so, they often disguise problem ground and it is only when a new extension is being built that the defect is exposed.

There was a typical case in my office a few years back, when a new extension was in the process of being built. Everything went according to plan until the builder began digging in the rear garden area and discovered the remains of an old pit. After he'd been digging for a while, he realized that the original house builder had decided that if he filled the pit with demolition rubble and old timber it would save him money. By doing so, he created a problem for the second builder because not only did he have to dig the new footings deeper than normal (normal strip footings cannot be built on filled ground), but he was forced to clear out the buried debris and remove it, which costs money.

The fact is, when dealing with a small domestic extension, the best that you can do is make an assessment of the likely ground conditions, design for those conditions, and advise your client that the foundations by their very nature are subject to review once construction work starts on site.

However, you do not have to make your assessment totally blind. Without a doubt, your client will have done some gardening and if he or she has dug down to any depth will know what type of soil is there. Other people in the street may have had extensions built and may tell you what the ground was like. If you know any local builders or the local BCO they may also be able to advise you. So, ask the following questions:

(a) What is the soil like – is it clay, sand, rock?
(b) Did the neighbours have any problems when they built their extension and how deep did they need to built their foundations?

Past experience will also guide you. I know that on my 'patch', unless information supplied by my client or observations on site lead me to believe otherwise, the ground is more than likely to be clay (there is a brickworks not far away).

FOUNDATION TYPES

There are several types of modern foundations for domestic construction and the details and sketches of which have been provided later.

(1) traditional strip
(2) deep strip
(3) raft foundation
(4) piled.

If foundation type 3 or 4 is needed (Building Control will tell you) then the designer will need the assistance of a structural engineer/suitable specialist. Although the tied footing or raft is beyond the scope of this book, I have included a sketch of a typical raft foundation for reference.

MORE ON TRADITIONAL STRIP FOUNDATIONS

A typical traditional strip footing is indicated in Figs 13.1 and 13.2. Study the sketches carefully because you will have to use similar details on your drawings (albeit at a smaller scale). I normally indicate a minimum width of 600 mm on my plans, as this size complies, for the most part, with the

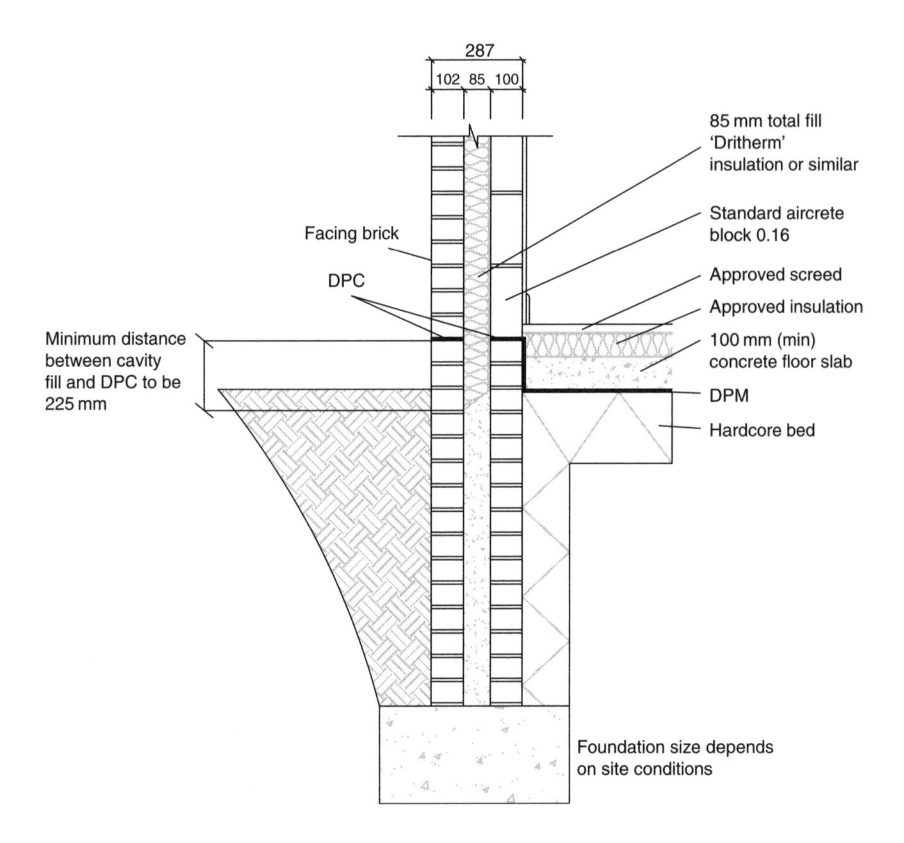

Fig. 13.1 Traditional strip footing: solid floors.

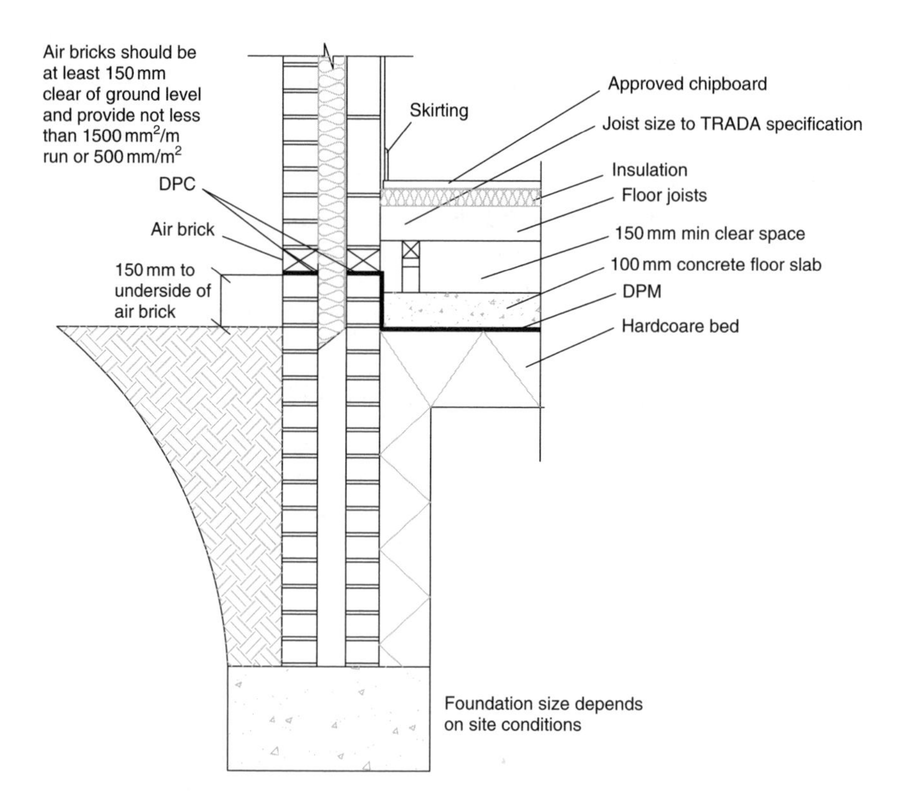

Fig. 13.2 Traditional strip footing: suspended floors.

requirements of Table 10 and Diagram 23 (of Approved Document A) and allows for most ground conditions. It is also a convenient size for a bricklayer to work off.

In order to be acceptable by Building Control Department, the traditional strip foundations must comply with the following:

(1) Consider the first three rules given under the section on 'Generally';
(2) The concrete must be an approved grade (e.g. GEN 1 or ST2). Although the minimum thickness of a foundation is 150 mm (6 in), where there are changes in level (stepped footing) the 'stepped' section must have minimum thickness of 300 mm. If piers are indicated on walls then the footing has to be enlarged to give the same projection. If you refer to Fig 13.1, you will note that I have indicated a 200 mm thickness as the foundation and not the 150 mm minimum. The reason for this is that if you take the wall thickness from the width of the concrete footing, the projection exceeds 150 mm. The thickness of the concrete has therefore to be greater than 150 mm and 200 mm is the next convenient thickness;

(3) Walls must be built centrally on a concrete footing. The BCO will reject your plans if walls are built 'off centre' unless you can prove that the non-traditional foundation is acceptable;

(4) If you refer to Chapter 8 (8.10), you will note that it is essential when considering foundation depths to take into account any trees growing in the area. Even though the tree(s) in question might not be in the garden of the property in question, it (they) might have an effect on the proposed design. On highly shrinkable clay soil, the NHBC set a minimum depth of 1 m as long as there are no trees present or recently removed from the area. Shallower foundations are allowed under NHBC regulations where the soil has a medium or low shrinkage potential. However, I have had it brought to my attention that in some parts of the country, because of a spate of subsidence problems, some local authorities are requiring foundation depths much deeper than 1 m. On my 'patch', one Local Authority Building Control Department normally encourages designers not to dimension foundations at all but merely state on the plans a clause along these lines of 'the foundations shall be of a size and taken to a depth as approved by the Local Authority Building Control Officers'. Unfortunately, clauses of that nature, whilst giving the BCO total control of that part of the works, do not help a tradesman builder who is trying to work out a quote for carrying out the works.

DEEP STRIP/TRENCH FILL FOUNDATIONS

These are similar to traditional strip footings and must comply with the same rules, but the concrete is deeper as shown in Fig. 13.3. When market conditions are favourable and when ready-mixed concrete is cheap to buy, many builders prefer to use trench fill footings because it reduces bricklaying time in the foundations, which speeds up the job. Deep strip/trench fill footings are also used where ground conditions dictate deep excavation or where the ground is influenced by trees.

THE RAFT FOUNDATION

As indicated above, I only intend to touch upon the tied footing/raft foundation because the subject is not covered by the Building Regulations and is really in the province of structural engineers. The reason for mentioning the tied footing/raft foundation is because it is a very useful form of foundation. The simplest form comprises a reinforced floor slab with a toe beam (Fig. 13.4) at the edge. Because it is outside normal building control parameters, calculations are required by most Building Control Departments before tied footings/rafts can be used.

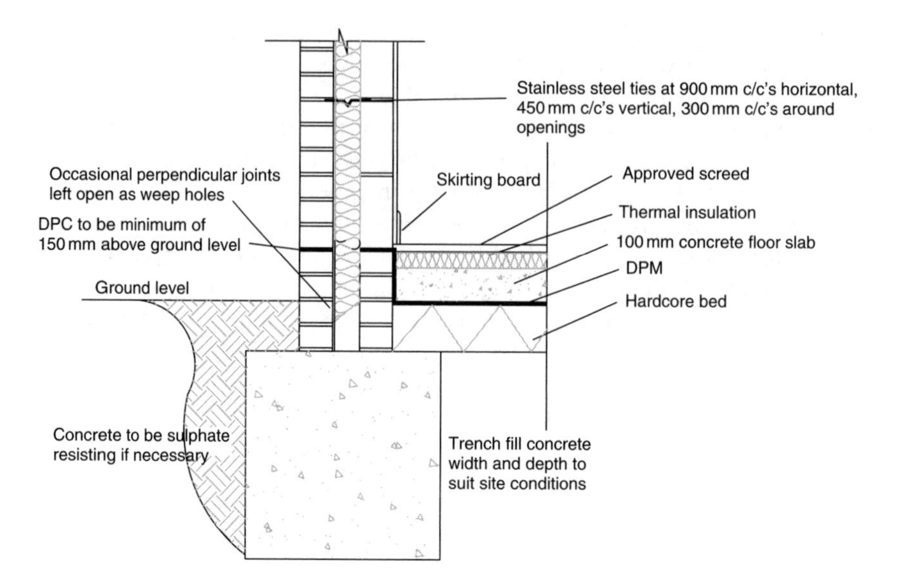

Fig. 13.3 Trench fill foundation.

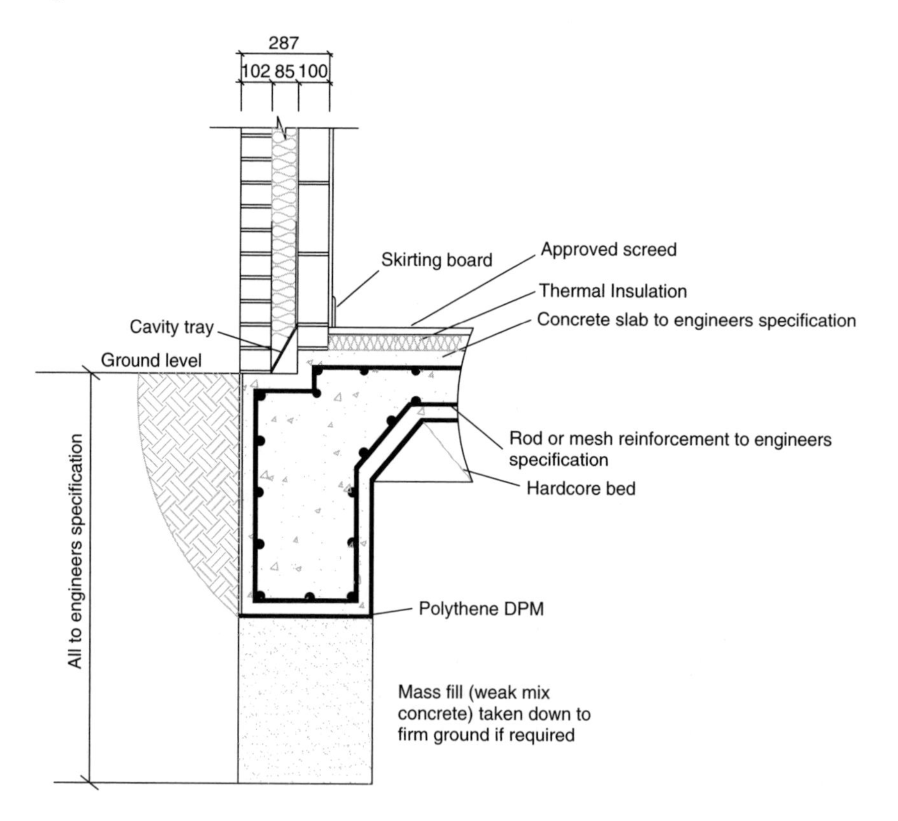

Fig. 13.4 Typical tied footing (raft foundation).

It is a useful form of construction where it is necessary to build up to a boundary line but the next door neighbour will not allow the spread of a normal footing to pass under their garden. I would suggest that it is always prudent to ask an engineer to prepare the raft calculations because most local authorities require an engineer's input before they will pass the plans.

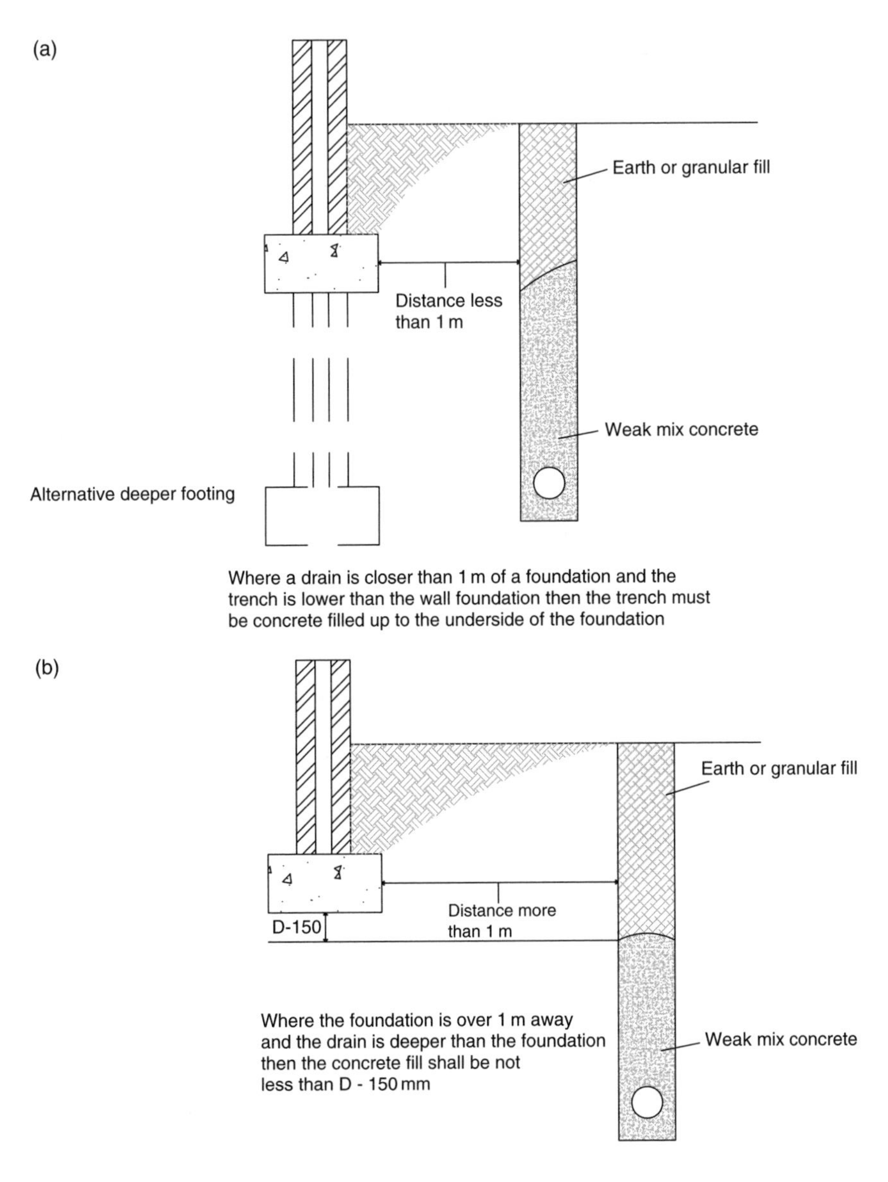

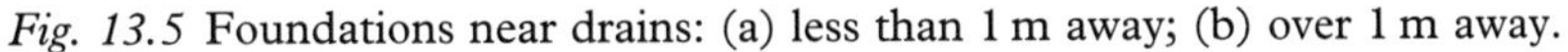
Fig. 13.5 Foundations near drains: (a) less than 1 m away; (b) over 1 m away.

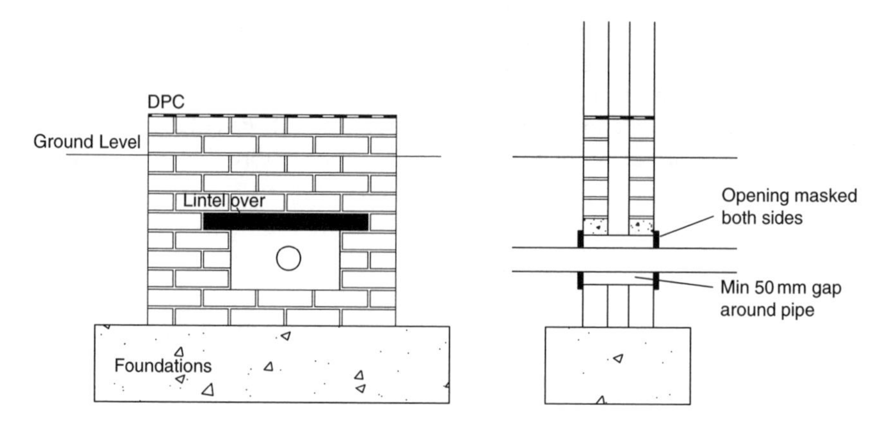

Fig. 13.6 Pipes passing through substructure walls.

DRAINS RUNNING NEAR OR THROUGH FOUNDATIONS

Sometimes existing drains on site can be deeper than the bottom of the proposed footing. The necessary measures to prevent collapse are indicated in Fig. 13.5. Where drain pipes have to pass through foundation walls they should pass through a properly formed hole with a lintel over it so that they are not cracked if the building settles slightly (Fig. 13.6).

14 The external walls

BRICK BONDING

Although timber frame construction (and SIPs – see later for more details) is now claiming a growing share of the housing market, brickwork is still a major element in most houses/extensions. All brick and block walls should be built to a recognized bond. On external walls built in facing brick or selected commons the evenness of the bond is essential for appearances sake. (On walls that have a rendered finish, the neatness of the bonding is not so essential but the basics of bonding must be adhered to for stability. See Figs 8.3 to 8.9.)

TRADITIONAL EXTERNAL WALLS GENERALLY

As indicated above, although timber frame construction is now starting to displace standard brickwork, most external walls to modern domestic properties built in England and Wales are still of cavity wall construction. Although there had been some experimentation with cavity construction during the nineteenth century, cavity walls became more common at the beginning of the twentieth century in an attempt to overcome one of the major problems being experienced with solid walls, namely damp penetration.

Because stretcher bond is simple to use with cavity walls, it is the form of bonding most commonly seen on the external walls of modern housing (see Fig. 8.3). However, there are other bonds (e.g. English bond (Fig. 8.4), English garden wall bond (Fig. 8.5) and Flemish bond (Fig. 8.6)). When you are carrying out your survey prior to preparing plans, note the brick bonding on the existing building. The bonds shown usually give a strong indication of the existing wall construction. Stretcher bond usually indicates that the walls are of cavity construction (but not always). English bond, Flemish bond and English garden wall bonds usually indicate solid wall construction. The subject of bond is discussed in greater detail in Chapter 8 (section on Matching materials).

Solid walls can still be used in modern construction, but now a water-repellent defence has to be installed (e.g. external tile hanging or external rendering). As the cavity wall is now the most common, I intend to concentrate on this form of construction.

A cavity wall comprises two distinct skins of walling tied together with wall ties. The theory of the cavity wall is that the outer skin will obviously become wet when it rains but if the water manages to saturate the outer skin and forces its way inwards, when it reaches the cavity, it will merely run down the inner face of the outer skin and not dampen the inner leaf. Sometimes, even this defensive barrier is breached when building operatives become lazy and do not maintain a good standard of workmanship (e.g. bricklayers not keeping the cavity clear of mortar droppings). However, despite the human factor, there can be no doubt that cavity walls are far superior to solid walling in terms of weather resistance.

For more basic information on cavity wall construction, refer to Chapter 6.1 of NHBC Standards. This provides a wealth of information concerning building construction and good practice.

Approved Document L1B (see Chapter 12 for details) of the Building Regulations covers the standards required in modern buildings to help conserve fuel and power. As traditional building materials cannot easily provide the sort of thermal values needed to save power, it has become standard practice to incorporate insulating materials into external walls. This can be done by using insulating blockwork in one or both other skins of the hollow wall or by incorporating other insulants in the structure of the walls. Alternatively, it is possible to use a combination of products.

In some instances the insulation is placed on the outside or on the inside of the external walls. However, doing this has its disadvantages (insulation tends to be thick and takes up valuable space). Although it was resisted for a long time, it was inevitable that the void in the cavity wall should be targeted as a convenient area where the insulant could be sited without causing the external wall to become excessively thick.

Figure 14.1 shows typical cavity walls (both total fill and partial fill). As the name implies, the total fill cavity wall has all the cavity filled with an approved insulant such as Knauf Insulations's 'Dritherm' which is impregnated to resist water penetration.

Note: In some parts of the UK with high rainfall, total fill can only be used if special precautions are taken. If you refer to NHBC Standards Chapter 6.1, Appendix 6.1-A this details suitable wall constructions for use with full cavity insulation.

The NHBC Standards also provides a map of the UK which provides an indication of the likely exposure to wind-driven rain. As one would anticipate, the map indicates that mountainous areas of the UK such as parts of Northern Ireland, the outer islands to the North and West of Scotland, the Isle of Man, Western Scotland, most of Wales, England's Lake District, the Pennines and Cornwall and Devon the driving rain index is classified as being 'Very Sever'. In these areas, under NHBC Standards, total fill walls have to be clad or adequately rendered. However, it is also

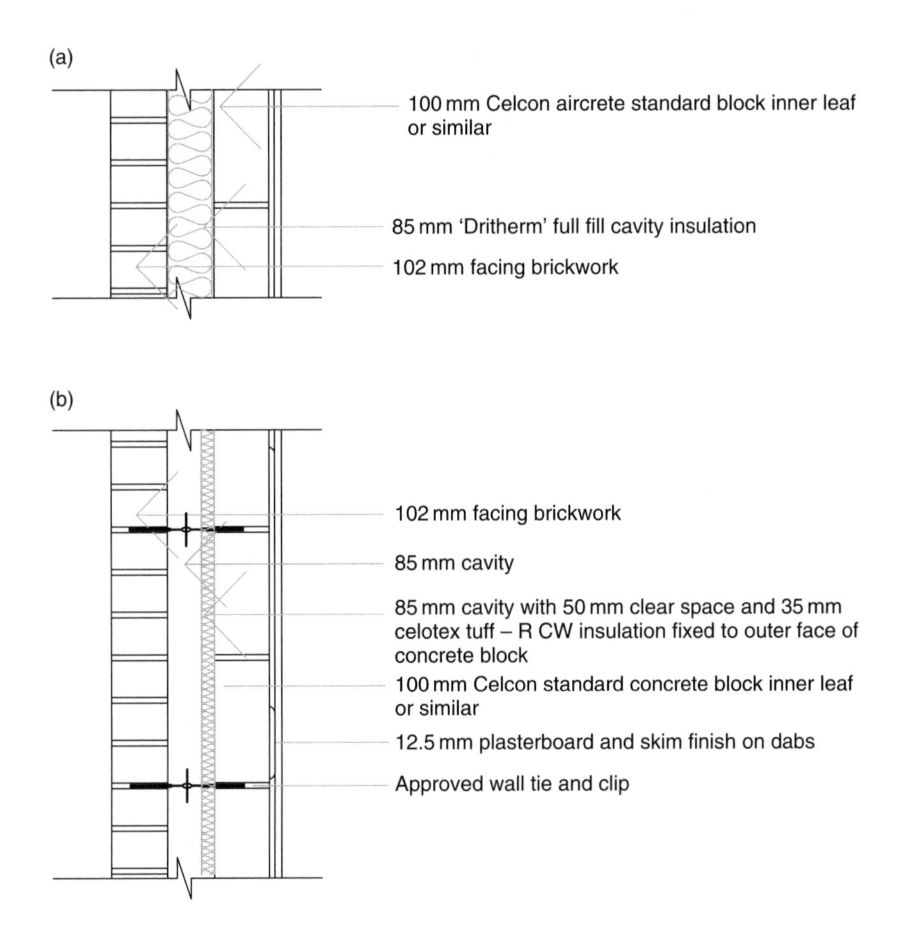

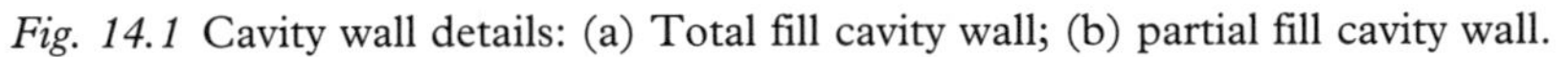
Fig. 14.1 Cavity wall details: (a) Total fill cavity wall; (b) partial fill cavity wall.

essential that you check with the specific manufacturer's recommendations and any locally applied Building Regulations.

Useful Weblink

http://www.knaufinsulation.co.uk/output/design/page_165.html

TIMBER FRAME CONSTRUCTION

There are other forms of external wall construction such as timber frame construction where the main load-bearing elements of the house are made in a factory. In many cases, a brickwork outer skin is then built around the factory-made inner units. Not only is the brickwork non-load bearing, it is really only provided to pander to the British need for good old bricks and mortar. Although timber frame construction is beyond the scope of this book, it is worth a mention because you may well become involved in the alteration of a timber-framed house. When you do, beware! In my experience, it is usually necessary to engage the help of a structural engineer before alterations are made to the timber framework. NHBC Standards Chapter 6.2 provides basic information on timber frame construction.

SIPs – USING STRUCTURALLY INSULATED PANELS IN THE WALLS OF AN EXTENSION

Structurally Insulated Panels (SIPs) consist of two outer boards of orientated strand board (OSB) sandwiching an inner core of Expanded Polyurethane MCFC Free, ODP ZERO.

Although comparatively new to the UK, SIPs have been used in other countries for a number of years. The advantage of SIPs is that when the building is engineered correctly and built with SIPs boards, no frame is required to support it.

SIPs can replace some of the traditional walling of an extension. The main element, which is replaced by using SIPs, would be the block work. One advantage of SIPs over block work is that SIPs can be erected in wet or cold weather. As SIPs come in widths of 1200 mm (4 ft) and 2400 mm (8 ft) they are quick to erect. According to SIPBuild, the core material has a ozone depletion rating of zero and the insulation benefits of the panel 150 mm thick is 0.17w/mk which is far better insulation standard than that currently required by the Building Regulations. It is also possible to achieve an airtightness of less than one air-change per hour compared to current targets of less than 10 per hour. All these benefits comply with the code for sustainable housing which has recently been published, with the aim of constructing a carbon neutral building to help achieve the reduction of the UK's 'carbon footprint'

The writer would like to thank Peter Barr of SIPBuild for his help, technical expertise and assistance in writing this section. (For further information dial SIPs telephone T:0870 850 2264.)

Useful Weblink

www.sipbuildltd.com

CAVITY WALLS BELOW GROUND LEVEL

The walls below ground level usually comprise two half-brick skins or two 100 mm blockwork skins or a combination of both. Alternatively, the two skins can be replaced with solid trench block footings. Where two skins are used (see Figs 13.1 to 13.3) a weak concrete cavity fill is inserted between the two skins to prevent them from being squashed together when building operatives backfill the trenches and destroy the stability of the foundations.

The weak concrete cavity fill must be kept down from the DPC by 225 mm. If the weak concrete cavity fill is shown higher than this, there is a danger that any water collecting at the bottom of the cavity during wet weather might breach the DPC defences in the wall, especially if the bricklayers have allowed 'snots' of mortar to fall down the cavities when they are building the superstructure.

CAVITY WALLS ABOVE DPC LEVEL

As indicated in Chapter 12, at the time of writing, external superstructure walls to habitable areas of a house should now have a *U*-value of 0.30 W/m^2K (the *U*-value is the thermal transmittance coefficient). Where cavity fill is used to attain the required *U*-value it has to comply with the requirements of Approved Document D of the Building Regulations and be inert and non-toxic (e.g. be an approved product such as 'Dritherm').

OTHER BUILDING CONTROL REQUIREMENTS – STABILITY OF EXTERNAL CAVITY WALLS

Approved Document A of the Building Regulations (STRUCTURE) is a vitally important document because it supplies design guidance for small buildings of traditional construction. I would suggest that you obtain a copy of Approved Document A and study it. (Copies of Approved Document A and the other Approved Documents can be obtained free of cost via the Internet.)

Although I have provided some basic information on Approved Document A, it must be borne in mind that it is not possible to condense all the rules contained within Approved Document A into a few pages and you must be prepared to study the whole document. The other thing to remember is that any set of rules is open to interpretation.

Listed below are *some* of the limitations that are stated in Approved Document A. If you exceed *any* of the limitations listed below (or any not listed below but contained within Approved Document A), then you will require the assistance of structural engineer to ensure the stability of the structure.

When designing walls/extensions/full houses, some of the most important points to bear in mind are

(a) You cannot use Approved Document A to design a skyscraper. On page 12 of Approved Document A – Section 2C (Thickness of walls in certain small buildings), it states that the information only covers residential buildings not more than three storeys high. (See Diagram 1 on page 13 of Approved Document A.)

(b) Most traditional external walls for small extensions are built using cavity construction (walls built in two leaves). There are minimum thicknesses of external walls. If you refer to Table 3 (page 15 of Approved Document A) you will note that the minimum thickness is determined by the height and length of the wall. In most cases the minimum thickness is 190 mm. However, as most small traditionally built home extensions have cavity walls which will be nearly 300 mm, compliance is virtually automatic.

(c) On page 13 of Approved Document A it states that all cavity walls must have internal skins at least 90 mm thick and 50 mm cavities. (**Note**: The details indicated in this book comply with these rules as they are shown with 100 mm inner skins and minimum 50 mm cavities.)

(d) Cavity walls (walls built in two leaves) need to be tied together with wall ties. See clauses 2C8 and 2C19 and Table 5 on page 21. The spacing of wall ties is also specified.

(e) Bricks and blocks come in a variety of compressive strengths. The compressive strength is described as N/mm^2 (Newton's per millimetre squared). It is important that you specify the correct type of brick and block for the project. If you refer to Diagram 9 on page 22, you will note that it indicates that when a traditional brick and block cavity wall is constructed the bricks shall be minimum $5\,N/mm^2$ for the brickwork element and $2.8\,N/mm^2$ for the blockwork. (**Note**: I tend to play safe and specify $4\,N/mm^2$ for the blockwork element to obviate the danger of overloading walls where beams bear on the blockwork.) In the lower storeys of three-storey extensions, stronger brickwork and blockwork are required.

(f) The building must not exceed 15 m in height (approx. 49 ft). This maximum is then qualified by Diagrams 6 and 7 of Approved Document A which requires location and local wind speeds to be factored in. However, as most domestic extensions do not normally come anywhere near 15 m in height, this calculation will probably not be required in most cases.

(g) The height of the building should not exceed twice the least width of the building. (See Diagram 1 on page 13 of Approved Document A.)

(h) Note the other requirements of Diagram 1.

(i) Note the requirements of Diagram 2 on page 14 of Approved Document A. Garages and other non-residential buildings and annexes also have size limitations. Flat-roofed annexes should not exceed 3 m in height (measured from the underside of ground slab). With pitched-roof annexes, the external wall height is 3 m (max), mid-gable 3.5 m (max) and 4.5 m max at an abutment with the main house (measured from the underside of ground slab). The maximum roof slope is 40 degrees.

(j) Floor spans – walls should not support any floor with a span larger than 6 m (measured centre to centre of bearings. See Diagram 11 on page 24).
(k) On page 17, Diagram 5 of Approved Document A, the maximum floor area of a room enclosed by structural walls is 70 m^2 or when open on one side 36 m^2. This is an important qualification because it places limits on room sizes.
(l) A thicker upper wall cannot be supported on a thinner lower wall. As an example, you cannot build a cavity wall upstairs on an existing half brick wall to, say, an existing garage downstairs. The existing garage has to be altered to provide cavity walls unless you have an engineer who can provide designs for, say, steel supports.
(m) Referring to Diagram 13 of Approved Document A (see also Fig. 14.2), where an internal partition wall acts as a buttress to an outer cavity wall, door openings in the partition have to be at least 550 mm away from the outer wall otherwise the partition will not be treated as a buttressing wall.
(n) You are advised to study Diagram 15 of Approved Document A because what I am about to state does not cover all circumstances, but in general terms, piers between windows and doors in external walls must be at least a sixth of the combined width of the openings. For example, if there are two windows in a wall each 1.20 m wide, the pier between them must be at least $(1.20+1.20)/6=0.4\,\text{m}=400\,\text{mm}$ (Fig. 14.3).
(o) Diagram 15 of Approved Document A indicates that an external wall cannot just end. It has to be supported by a buttressing wall. The minimum length must be 665 mm (or an engineer's calculation has to be provided). **Note**: The 665 mm rule does not apply to garages which are covered by Diagram 18 of Approved Document A.

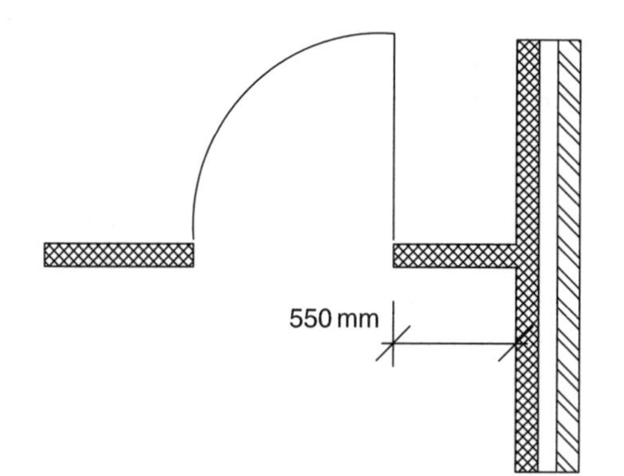

Fig. 14.2 Door opening.

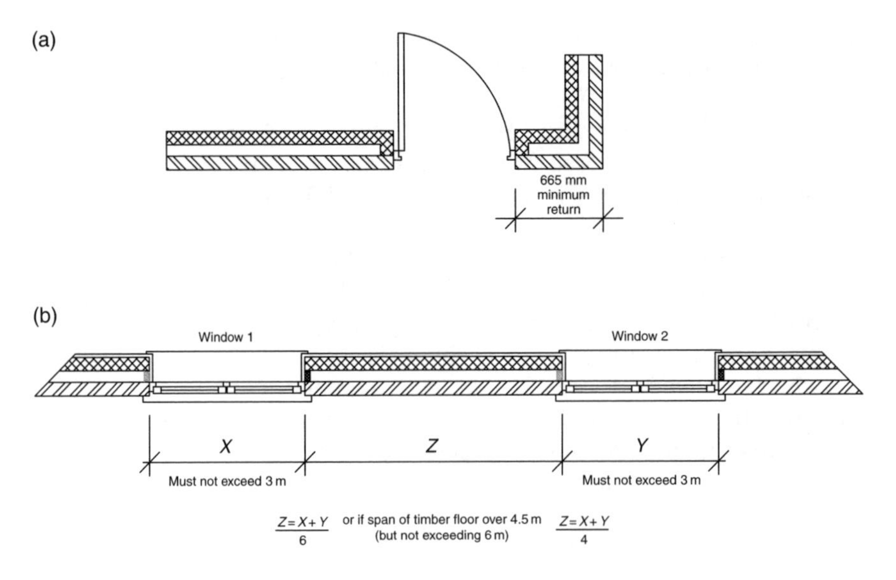

Fig. 14.3 (a) External corner; (b) pier between openings. (Details after approved document A.)

Useful Weblink

http://www.communities.gov.uk/

THE USE OF RESTRAINT STRAPS/LATERAL SUPPORT

In older properties, bulges in external walls are very often noticeable. Although corroding cavity ties can sometime be to blame, lack of lateral support is another cause. In theory, the floor and ceiling joists of a house act as ties and hold the walls in position but the number of failures has proven that the walls do move away from the joists, especially where walls run parallel to the joists or where a staircase runs alongside a wall.

The solution to the problem is the incorporation of metal restraint straps which ensure that walls, floors and roofs are secured together to prevent movement. Pages 29 and 30 of Approved Document A detail the requirements for lateral support. The basic principle is that flat roofs, first floor joists, ceiling joists and roof timbers should be strapped to the walls at centres not exceeding 2 m with galvanized mild steel (stainless steel is better but very expensive), 30 mm × 5 mm cross section. Where joists run parallel to the wall, the restraint straps run at right angles across the joists.

LINTELS AND BEAMS

The most usual type of lintel used on modern extensions is metal 'Catnic' or similar type (Fig. 14.4). These lintels/beams are made from sheet steel

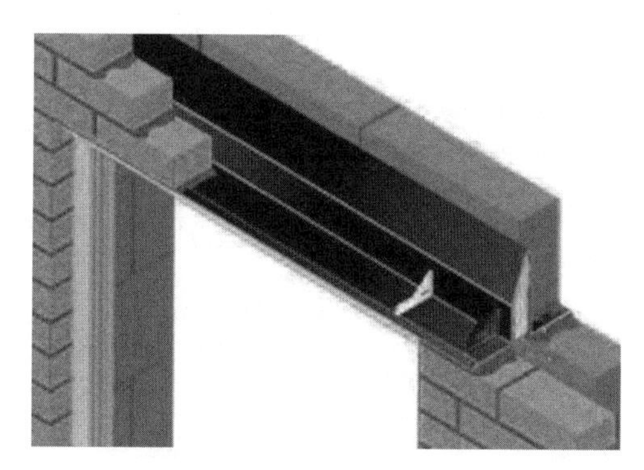

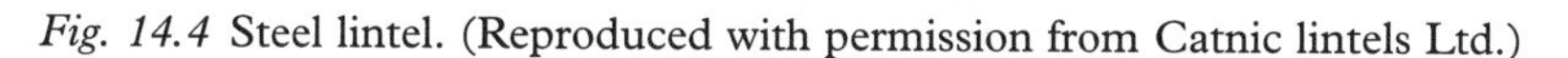

Fig. 14.4 Steel lintel. (Reproduced with permission from Catnic lintels Ltd.)

that has been galvanized or treated against rusting, come in a wide variety of forms and have been designed by the manufacturer to take normal loads above door and window openings. If you are in any doubt about the strength of any lintels specified, most manufacturers will check the loads for free of charge if you send them a copy of your plan (make sure that there is no charge prior to sending the plan because company policies can change from time to time). To comply with thermal regulations, most lintels now have insulation incorporated into them.

Manufacturer's regulations should be followed, but the usual bearing for a lintel is 150 mm (6 inches) each end. If you don't give enough end bearing, the lintel will fail under load.

The advantage to the designer of using any of the standard lintel/beam types is that Building Control will not usually require calculations to prove the lintels (unlike RSJs, see below for details) as long as the manufacturer's load tables are followed. Although I have provided some information on these types of lintels, try to obtain manufacturers' catalogues and study the details in greater depth.

Where one is forming an opening in an existing wall to connect the new extension to the existing, it is normal practice to install two RSJs (rolled steel joists) or steel beams bolted together and seated on adequate concrete padstones. The 'Catnic' type beam can also be specified, but the reason why I normally avoid patent beams for inserting into old walls is simple practicality. Any beam being inserted into an existing wall tends to be 'bashed about a bit'. It would be difficult for even the heaviest handed building operative to damage what is still commonly known as a RSJ (see Fig. 14.5).

Those having no experience with beam calculations can best obtain them from a structural engineer. Those with some basic knowledge can use one of a number of simple and (and inexpensive) steel calculation programs which are now on the market.

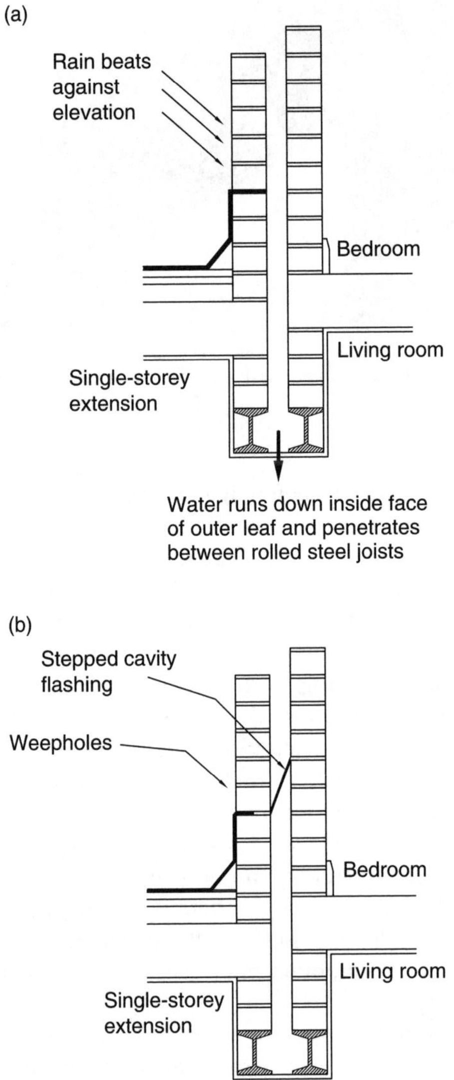

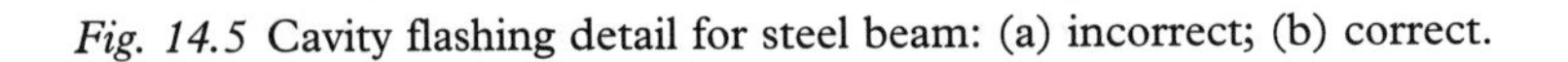

Fig. 14.5 Cavity flashing detail for steel beam: (a) incorrect; (b) correct.

Useful Weblink

http://www.corusconstruction.com/en/products_and_services/building_products/catnic/products/lintels/

15 The internal walls

PARTITIONS GENERALLY

In a modern house the upstairs partitions are usually what is known as stud partitions (a timber frame with plasterboard both sides).

However, as ground floor partitions often provide support for the floors above, some of the ground floor partitions are normally made from blockwork. It was common practice in the not too distant past to build load-bearing internal blockwork partitions off the floor slab and, other than a slight thickening in the slab, not to bother with a proper foundation.

BLOCKWORK PARTITIONS

As blockwork is usually far heavier than studwork, and as other loads from the roof and upper floors were being transmitted down onto the lower floor slab, the inevitable result of this practice was that cracks developed. In bad cases of overloading, the internal walls also developed cracks as the weakened ground floor slab gave way.

In my experience, nowadays, most Building Control Departments query the use of slab thickening (unless engineers' calculations are supplied) and require the blockwork partitions (whether load-bearing or otherwise) to be taken down to a proper foundation similar to the external walls. NHBC Standards also require the same.

Openings in block partitions

Nowadays it is normal to incorporate a standard lintel such as a Catnic CN102. You must however check with the manufacturers handbook to ensure that the lintel that you are using is suitable for the location.

Wider opening may require steelwork inserting. NHBC Standards Chapter 6.5 provides a schedule of suitable steel beams capable of supporting various partitions. If you do use these details, it will probably be necessary to submit extracts from the NHBC Standards with your building control application to 'prove' your design.

Note: These beams are not designed to take floor loads. If the beam needs to take floor and wall loads refer back to comments made in Chapter 14 concerning lintels and beams.

Useful Weblink

http://www.corusconstruction.com/en/products_and_services/building_products/catnic/products/lintels/

THE BASIC PURPOSES OF PARTITIONS

Ignoring fire/smoke protection, the internal walls/partitions of a house serve four main purposes:

(a) To subdivide the property into separate rooms.
(b) To provide support for the roof.
(c) In a semidetached or terraced property, to separate the dwelling from its neighbour and provide fire and sound protection between dwellings.
(d) Provide a reasonable resistance to the passage of sound: between bedrooms and other rooms/internal walls between a toilet and other rooms.

In dealing with internal partitions (c) and (d) have now become the most important. In previous chapters there has been a distinct emphasis on

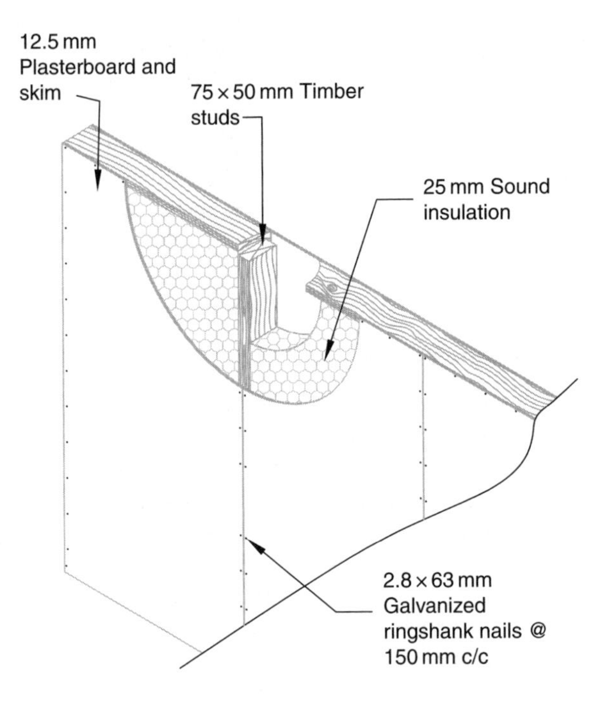

Fig. 15.1 Timber stud partition.

preventing heat loss. However, when dealing with internal walls/partitions this changes. Heat insulation is virtually ignored and sound insulation becomes the big issue.

The reason for this is that uncontrolled noise transmission is one of the biggest headaches of modern society. Someone playing loud music late at night can cause real aggravation to another party.

THE EFFECTS OF APPROVED DOCUMENT E – RESISTANCE TO THE PASSAGE OF SOUND

At one time, builders paid little attention to the passage of sound. Because they were relatively cheap, upstairs partitions were generally nothing more than timber framework covered with a layer of plasterboard each side and noises in one room very easily traveled to another.

The 2003 edition (with 2004 amendments) of Approved Document E has changed that. In a typical domestic situation, stud walls around bedrooms and bathrooms need to be minimum 75 mm studwork with either two layers of plasterboard on each side or the partition has to incorporate 25 mm of sound insulation within the wall (see Fig. 15.1).

For extensions where more than one property is involved, the separating walls should be built to the standards required for new dwellings. Section 2 of Approved Document E provides guidance by providing four wall types that will satisfy the regulations.

Type 1 construction comprises (ignoring the concrete wall) either:

(1) a 215 mm solid masonry wall with a brick density of 1610 kg/m^2 (110 mm coursing)
(2) or a 215 mm block laid flat with a block density of 1840 kg/m (75 mm coursing).

with 13 mm lightweight plaster (minimum mass per unit area 10 kg/m^2).

In this sort of situation, the new internal party wall will intersect with the cavity wall and a cavity stop will need to be inserted as per Diagram 2.5. (**Note**: The cavity stop will not be required if a total cavity fill solution such as Dritherm is adopted.)

16 Ground floors

GENERALLY

Ground floors to domestic extensions can be divided into three basic types:

(1) ground supported solid floors (see Figs 13.1, 16.1 and 16.2)
(2) suspended timber floors (see Figs 13.2 and 16.3)
(3) patent suspended concrete ground floors (e.g. beam and block) – As this is a patent form of construction, individual manufacturer's catalogues need to be consulted. The type of floor chosen depends upon the designer's preference and site conditions.

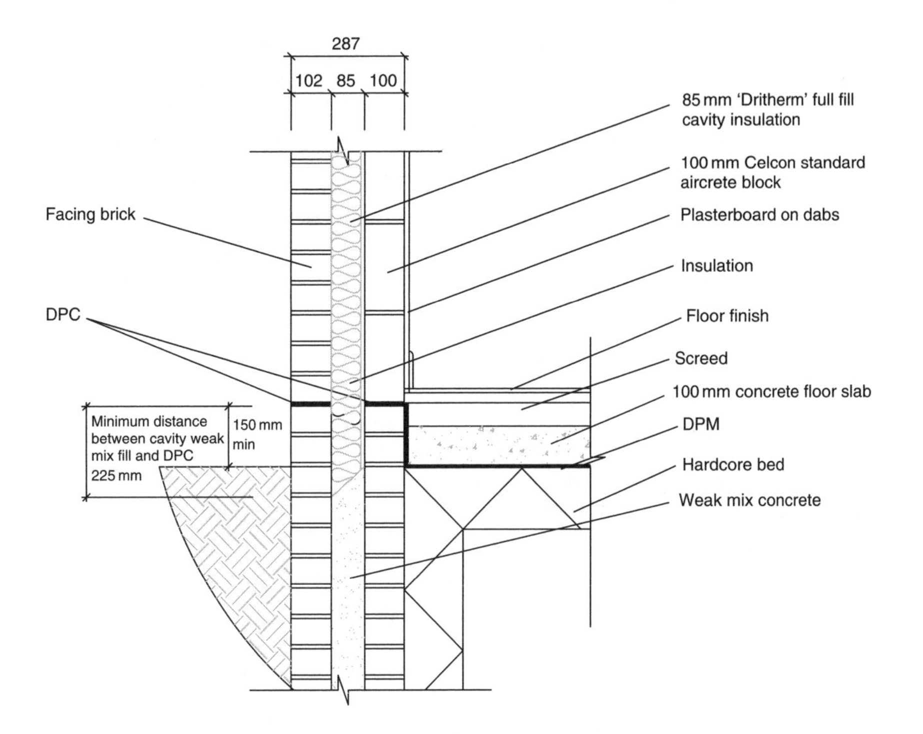

Fig. 16.1 Uninsulated solid ground floor.

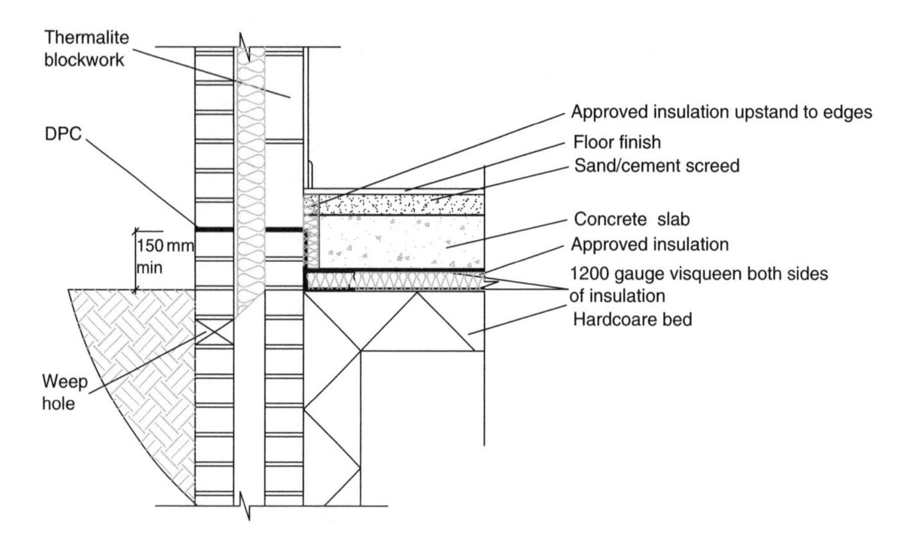

Fig. 16.2 Insulated solid ground floor.

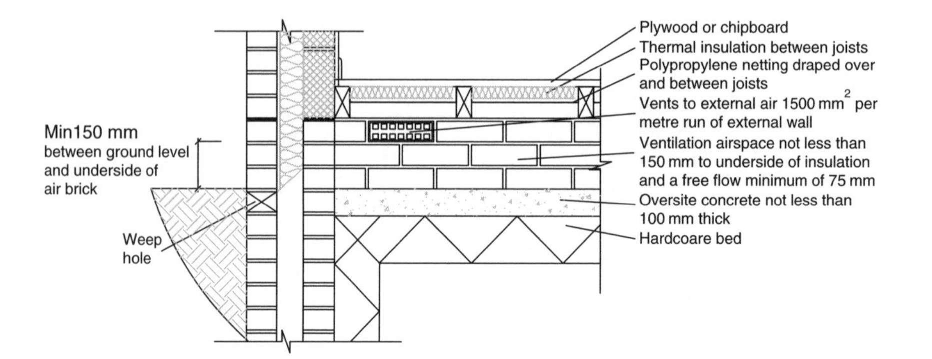

Fig. 16.3 Suspended timber ground floor.

Useful Weblinks

http://www.bison.co.uk/Beam-and-block-floors.aspx
http://www.knaufinsulation.co.uk/output/solutions/page_233.html

COMMON FAULTS WHEN DEPICTING THE GROUND FLOOR CONSTRUCTION ON A DRAWING

When preparing a plan for an extension/dwelling the following details must be observed:

Do not show the surface level of finished ground floor slab or subfloor surfaces as being lower than external ground level. Flooding is becoming a major problem in some parts of the UK. If the floor slab or subfloor is built too low, the extension could flood in wet weather.

GROUND SUPPORTED SOLID FLOORS

The requirements of Approved Document C

Section 4 of Approved Document C of the Building Regulations provides the basic requirements for ground floors (e.g. Floors next to the ground).

As indicated earlier, all the Approved Documents can be downloaded from the Internet free of charge so I would suggest that you download a copy and go through the items below.

Clause 4.2 of the Approved Document requires solid ground floors to comply with the following:

(1) Ground floors must be able to resist the passage of moisture. (Translation – Incorporate a 1200 gauge damp-proof membrane or DPM.)
(2) The floor should be made of a substance that won't be damaged by ground water.
(3) The floor should resist the passage of gases. (In some areas methane or radon can be in the ground upon which the extension is built. In these cases, a gas membrane has to be installed or other solution acceptable to Building Control.)

Clause 4.7 and Diagram 4 provide a technical solution for the majority of solid floor construction, namely:

(1) An approved hardcore bed (with a sand blinding – The sand blinding is to prevent the polythene DPM from being punctured). (**Note**: The hardcore must not exceed 600 mm in thickness. If it does another form of construction must be employed.)
(2) A DPM (Translation – 1200 gauge polythene sheet) laid over the hardcore but under the concrete slab and turned up and linked to the damp course in the walls.
(3) Minimum 100 mm concrete BS 8500 mix ST2 or ST4.

GROUND SUPPORTED SOLID FLOORS – UNINSULATED TYPE

Figure 16.1 depicts a standard uninsulated solid ground-bearing floor slab. As Approved Document L1B requires most floor slabs to be insulated, the construction shown in Fig. 16.1 would probably only be used in a building control exempted structure such as a conservatory.

GROUND SUPPORTED SOLID FLOORS – INSULATED TYPE

It is now a requirement of Approved Document L1B that most floors have a *U*-Value of 0.22 w/m^2 (see Approved Document L1B Table 4).

If you refer to Fig. 13.1, this shows a solid ground floor slab with the insulation above the slab. Figure 16.3 shows the insulation below the slab. Personally, I prefer the 'below slab solution' because it occupies 'dead space' and also helps to ensure that the hardcore does not exceed 600 mm in thickness. Most builders appear to prefer underfloor insulation.

If you read the Approved Document L1B, the correct way of calculating the thickness of insulation is by working out the perimeter/floor area factor. However, because life is too short to peel a grape, most designers work on an insulation thickness that will suit all cases. At the time of writing, the following appear to satisfy the criteria:

- 75 mm Polyfoam floor board
- 90 mm Celotex tuff-R™ GA3000 and Celotex T-Break™ TB3000 high performance thermal
- 80 mm Kingspan Kooltherm K3 floor board
- a word of caution! Make sure that you follow the insulation manufacturers instructions.

Useful Weblinks

http://www.knaufinsulation.co.uk/
http://www.celotex.co.uk/floors.php
http://www.insulateonline.com/index1.htm?floorsintro.htm~main

SUSPENDED TIMBER FLOORS

The requirements of Approved Document C

Clause 4.13 of the Approved Document requires suspended timber ground floors to comply with the following:

(1) The ground under the floor is covered to resist moisture and prevent plant growth.
(2) There is an adequate ventilated air space between the ground covering and the timber.
(3) There is a DPC between the timber and any material which can carry moisture from the ground.

Clause 4.14, and Diagrams 5 and 6 provide a technical solution for the majority of suspended timber ground floor construction, namely:

(1) It has a ground covering of at least 100 mm thick mix ST1 to BS 8500 laid on approved hardcore, or
50 mm inert fine aggregate (sand) laid on 1200 gauge polythene.
(2) Damp courses and ventilation as detailed on Fig. 16.3.
(3) Under kitchens/bathrooms and utility rooms where water can be spilled, the flooring should be waterproof grade (e.g. Type P5 moisture-resistant chipboard).

Section 4 of Approved Document C of the Building Regulations provides the basic requirements for ground floors (e.g. Floors next to the ground).

The timber ground floor (see Fig. 16.3) still requires

(a) A hardcore bed and 100 mm concrete slab as with the solid floor, or
(b) 50 mm of concrete or inert fine aggregate and a polythene DPM. This alternative has only very limited use. If honeycomb sleeper walls are needed (see below for details) option (a) should be adopted.

SUSPENDED TIMBER FLOORS – OTHER MATTERS

If you refer to Fig. 16.3 you will see a note that says honeycomb sleeper walls. Honeycomb sleeper walls are normally spaced at 1.80–2 m centres and are used to support the ground floor joists which reduces the spans and cross sections of joists being used in the floor. (**Note**: Honeycomb sleeper walls are walls built with air holes in them to provide good through ventilation beneath the floor.)

Obviously, where honeycomb sleeper walls are required, it is necessary to use a 100 mm underground slab indicated above.

Between the honeycomb walls and the joist, it is normal to incorporate a wall plate which is bedded on a DPC. When drawing your plan, keep the concrete subfloor surface above ground level.

Approved Document L1B also applies to suspended timber floors and insulation has to be provided. At the time of writing, the following appear to satisfy Building Control criteria:

- 140 mm Celotex tuff-R™ GA3000 and Celotex
  Extra-R™ XR3000 high
- 80 mm Kingspan Thermafloor TF73 zero ODP composite floor insulation.

The voids underneath timber floors have to be provided with ventilation to the outside air. This is done by installing air bricks, but they should not be built in too low, otherwise water will flood into the house in wet weather.

The purpose of air bricks beneath a timber floor is to prevent dry rot spores germinating on the wood of the floors. The ventilation should be equal to 1500 mm^2 per metre run of external wall or 500 mm^2/m^2.

Note: Air bricks will not work properly unless they are properly spaced so that they create cross ventilation. Bad ventilation can allow dry rot to infect the timbers.

Note on dry rot

Dry rot is caused by a fungus called *Serpula lacrimans*. Dry rot is the most serious timber decay problem in buildings in temperate regions. This fungus has the unique ability of being able to penetrate the non-timber elements of buildings, such as masonry and plaster, in the form of mycelial strands, and to transport water through those strands, allows the fungus to spread considerable distances from its point of origin.

However, to germinate, true dry rot needs three things:

(a) moisture in the wood over 20 per cent
(b) bad ventilation
(c) warmth.

This is why air bricks are so important. They provide an air flow which dry rot doesn't like, they dry out the floor timbers and prevent it from being available to the fungus and they reduce the temperature slightly.

As you will have already realized, there is a duplication of materials below a timber ground floor, which is why this form of construction is more expensive than a solid ground floor under normal conditions. On the other hand, if a large amount of fill is required below a solid floor because of steeply sloping site levels, it could be cheaper to put in a timber floor.

PATENT SUSPENDED FLOORS

In recent years, the 'Housefloor' system has been introduced in the UK and comprises purpose-made concrete joists infilled with standard building blocks. For details see Websites given above.

17 Timber upper floors

OBTAINING SIZES OF CONVENTIONAL FLOOR JOISTS

One of the most fundamental elements of floor design is deciding what size joists are needed to make the floor structurally sound.

Unlike suspended timber ground floors, the joists of upper floors have to be of a much larger section because the spans cannot be broken by means of such things as sleeper walls. The size of roof joists and the grade of timber required can be obtained from 'Span tables for solid timber members in floor, ceilings and roofs for dwellings' from TRADA Technology.

Useful Weblink

http://www.trada.co.uk

Alternatively, refer to NHBC Standards Chapter 6.4.

When you look at the tables, you will note that they are divided into dead load bands. I use the more than 0.50 n e 1.25 kN/m^2 tables because BCOs checking the submitted plans always err on the side of caution.

The tables are also divided into various grades(e.g. C16, C24); the higher the number the better the quality of timber and the larger the span but the more expensive. Try to stick with C16 unless you are forced to do otherwise.

While you look at Chapter 6.4, also study the other NHBC details because this will increase your knowledge of basic construction.

AVOIDING A BEGINNERS MISTAKE

When designing, say, a large two-storey extension with several rooms at ground level the temptation for most beginners is to choose floor joists of different sizes to suit each room. If you do this, and the builder doesn't query the instructions given on your plan, either the floor surface or the ceilings in the rooms below will go up and down.

Think about it!

The sensible solution is to find the longest span, select the joist size for the longest span and use the same size floor joist for the whole floor area.

Strapping floor joists to walls

If you refer to Fig. 17.1, you will note that the floor joists should have restraint straps fixed to them to provide lateral support to external walls.

Reducing spans by inserting steel beams

If you follow the advice given above, it will serve you well 95 per cent of the time. Unfortunately, there is a slight flaw in the advice. Timber is very expensive and by selecting the largest joist size, this could add a great deal of cost to the project. In addition, if you refer to the TRADA or NHBC timber tables, you will note that using the more than 0.50 n e 1.25 kN/m^2 tables the maximum span for standard timber joists is less than 5 m.

One way around both problems is to insert a steel beam at an appropriate location.

But what size steel beam?

NHBC Standards Chapter 6.5 provides a schedule of suitable steel beams capable of supporting various floor areas. If you do use these details, it will probably be necessary to quote/submit extracts from the NHBC Standards with your building control application to 'prove' your design.

Figure 17.2 indicates how a steel beam should be fitted to a timber floor.

I-Joists and metal web joists

While you are looking at NHBC Standards 6.4, note in particular the references to modern I-Joists and metal web joists. Although they are beyond the scope of this book, I-Joists and metal web joists are worthy of a mention

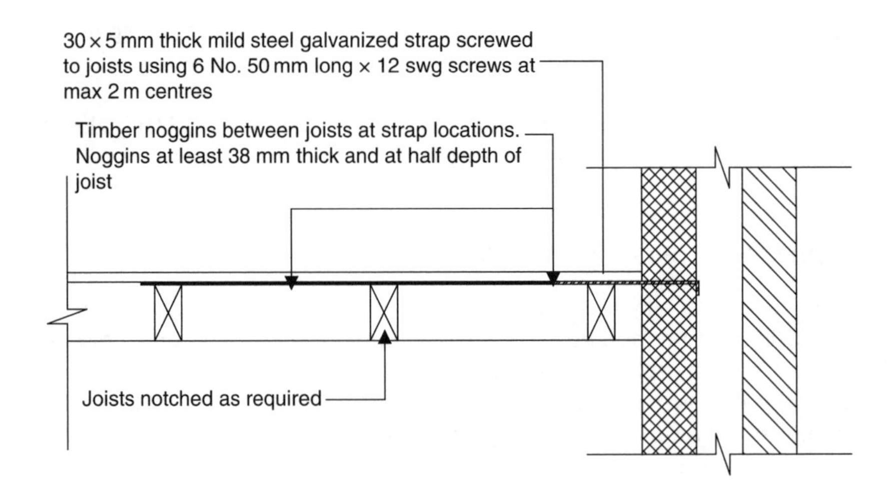

Fig. 17.1 Restraint straps with joists parallel to wall.

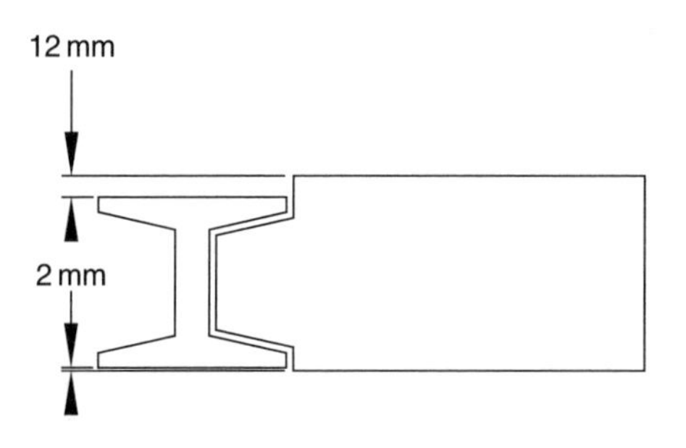

Fig. 17.2 Joists trimmed into steelwork.

because they will provide far larger spans than conventional timber floor joists.

If you do need to use I-Joists and metal web joists on a project, it will be necessary to either submit manufacturer's calculations/data sheets with the application or ask for conditional approval for them.

Note: Conditional approval for certain elements of structure can be requested when submitting the building control application. This is very common when using roof trusses because roof truss manufacturers will normally not supply calculations until they have received an official order of the builder. Obviously, the designer is not usually in a position where he can place an order.

OTHER BUILDING REGULATION REQUIREMENTS

Figures 17.3 and 17.4 show the general construction of timber upper floors.

Until recently, most upper floors in houses were not insulated. However, because of public complaints concerning noise pollution, Approved Document E (Resistance to passage of sound) requires that upstairs timber floors of houses are now sound insulated.

The requirements of Approved Document E

On page 66 of Approved Document E clause 5.23 and Diagram 5.7 provide deemed to satisfy details for the sound insulation required for an upper floor. The boarding (timber of wood-based board) has to have a minimum mass of 15 kg/m^2, the plasterboard ceiling has to have a mass of 10 kg/m^2 and there has to be an absorbant layer of mineral wool minimum density of 10 kg/m^3.

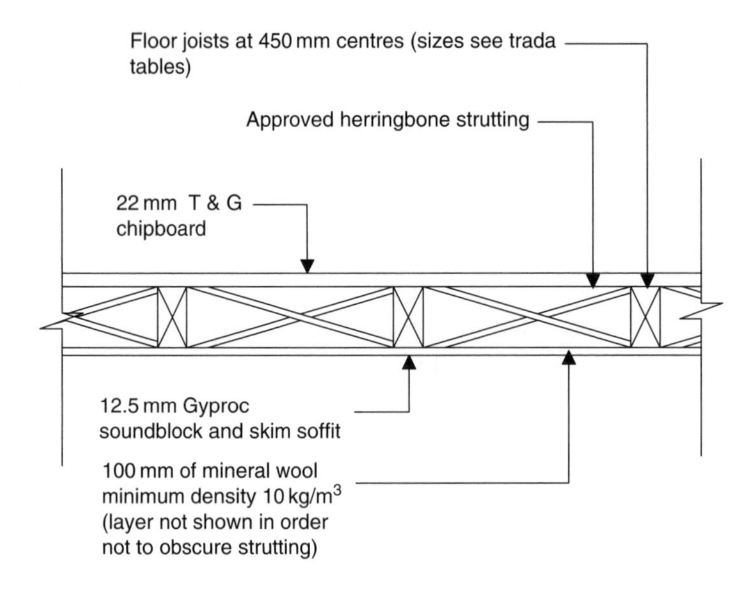

Fig. 17.3 Cross section of timber – joisted first floor.

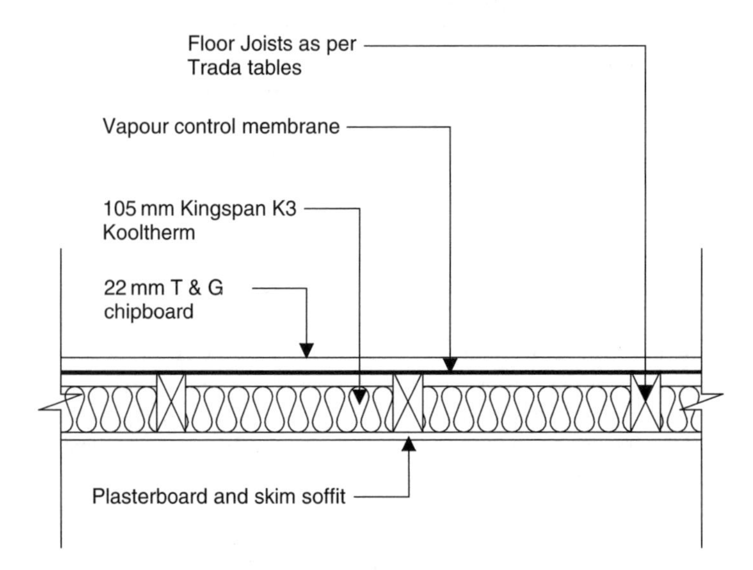

Fig. 17.4 Cross section of timber floor over partially ventilated space (e.g. garage).

Translated into simple English this means that the upper floor of two-storey extension should have a minimum the 22 mm tongued and grooved chipboard. The tongues and grooves should be glued with an approved glue. NHBC Standards recommend the use of Type P5 moisture-resistant chipboard. Inside the floor, there should be 100 mm mineral wool with a density of 10.5 kg/m^2. One other requirement is that the plasterboard ceiling needs to be at least 10 kg/m^2. One way of complying is to line the ceiling with Gyproc Wallboard 10 (or similar), which is designed to comply with the regulations.

Useful Weblinks

http://www.knaufinsulation.co.uk/output/design/page_165.html
http://www.british-gypsum.bpb.com/products/plasterboard_accessories/gyproc_acoustic/gyproc_wallboard_ten.aspx

The requirements of Approved Document L

Approved Document L requires floors to have a *U*-value of 0.22 W/m^2 K. If you follow the Approved Documents slavishly, the way to work out the thickness of insulation required is by using the perimeter to floor ratio. The other alternative is to establish the 'worst case' scenario; Using Celotex tuff-R™ GA3000 and Celotex Extra-R™ XR3000 high performance thermal insulation, 130 mm would be needed.

Useful Weblink

http://www.celotex.co.uk/floors/suspended.php

18 Flat roofs

GENERALLY

In the 1960s, a large number of home extensions had flat roofs. This type of roof generally comprised a timber or chipboard 'deck' overlain with three layers of felt bonded together with bitumen.

This type of roof is nowadays known as the cold deck flat roof and is highly prone to internal condensation problems. Water vapour condenses inside the flat roof void staining the plasterboard ceiling and dampening the flat roof joists and making them prone to rotting.

It is worth noting that cold deck flat roofs are no longer recommended by NHBC Standards.

The reasons given are the difficulty in providing

(1) effective vapour control layer at ceiling level
(2) required level of ventilation
(3) unobructed 50 mm ventilation over insulation
(4) ventilation at both ends of each joist void.

It is also worth noting that Approved Document C of the Building Regulations now indicates that flat roofs should be designed in accordance with Section 8.4 of BS 5250: 2002. In turn, BS 5250 indicates that because of the inherent condensation problems suffered by cold deck flat roofs, this form of construction should be avoided.

Although the term 'flat roof' is accepted terminology, flat roofs should not be totally flat, they should have a slight fall on them so that they drain. Unfortunately, some builders took the name literally and built them totally flat. Needless to say, in a very short time, they began to leak.

So why build flat roofs if they are prone to problems?

There were several reasons for the initial popularity of flat roofs. Although flat roofs have a short lifespan, they are cheaper to build than a tiled or slated roofs. However, it wasn't long before bad design (lack of drainage and internal condensation) and the short lifespan of the felts gave them a bad name.

To be fair to the flat roof industry, felt manufacturers have improved their products and supply high tensile felts which will last much longer than the old-fashioned felts.

However, mud sticks, and nowadays very few people ask for a flat roofed extensions any more. More importantly, some Local Authority Planning Departments have adopted anti-flat roof policies and resist the construction of flat roofs on domestic extensions unless there is no alternative.

If you have to use flat roof construction, opt for what is known as a warm deck flat roof.

THE WARM DECK FLAT ROOF

NHBC Standards Chapter 7.1 Appendix 7.1A depicts various types of warm deck. In Appendix 7.1B various types of approved roof finish are listed.

With the warm deck, the insulation is placed on top of the roof deck and the void between the deck and the plasterboard ceiling stays warm because it is protected by the insulation. On a warm deck roof, ventilation should not be provided to the enclosed voids.

WARM DECK STRUCTURE

Figure 18.1 depicts the main elements of the warm deck flat roof, starting underneath and moving upwards:

Plasterboard/plaster soffit

Like most modern ceilings, a flat roof is normally lined with plasterboard. As the joists will probably be at 400 mm centres, it will only be necessary to use 9.5 mm plasterboard with a skim finish of plaster. Then there are the roof joists. As with the floor joists, the size of roof joists and the grade of timber required for a flat roof can be obtained from 'Span tables for solid timber members in floor, ceilings and roofs for dwellings' from TRADA Technology. It would be sensible to specify that the roof timbers are treated using pressure impregnation with a water-based preservative, complying with BS 4072.

Firrings

Falls (slopes) are created by installing firrings on top of the roof joists (which should be laid horizontally) and below the plywood or chipboard roof deck.

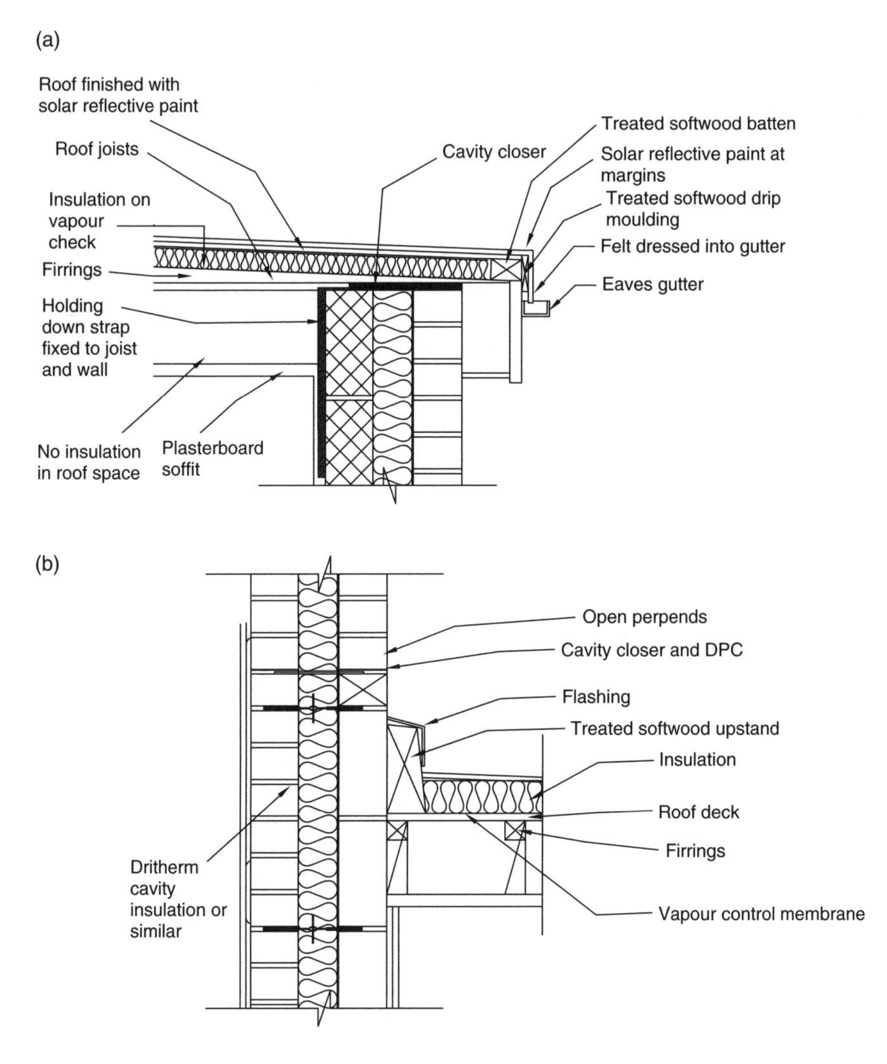

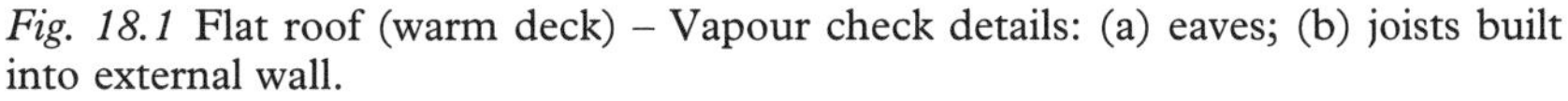
Fig. 18.1 Flat roof (warm deck) – Vapour check details: (a) eaves; (b) joists built into external wall.

Firrings are shaped timber battens that start wide at one end and then taper down uniformly to create a regular fall and/or crossfall on the roof deck. The NHBC recommend the following minimum (measured at the thinnest end) sizes of firring:

- joists at 450 mm centres or below – 38 mm × 38 mm
- joists above 450 mm up to 600 mm centres – 38 mm × 50 mm deep.

The roof deck

Supporting the finish there is the roof deck 15 mm pre-treated weather and boil proof (WBP) or marine plywood.

Roof straps

In recent hurricane force winds that swept the UK, some flat roofs literally took off and ended up in the back garden. In order to combat wind lift, it is now a requirement of the Building Regulations that all roofs are strapped down to the surrounding walls using 30 mm × 5 mm galvanized mild steel straps at 2 m centres.

Roof finishes

Appendix B provides the specification for flat roofing. The builder shall comply with NHBC Standards Chapter 7.1.

Gutterwork and downpipes

Approved Document H of the Building Regulations sets out the rules for the size of gutters and downpipes. Assuming only one downpipe and one length of guttering the following sizes are needed:

(a) For roofs of 18 m^2 the gutter has to be 75 mm (mm.) with an outlet of 50 mm (mm.).
(b) For roofs of between 18 and 37 m^2 the gutter has to be 100 mm (mm.) with 63 mm (mm.) outlet.

All rainwater pipes should discharge into gullies and from there to underground drainage. In order to ensure that water does not splash up from the gullies most builders connect the downpipes into back inlet gullies (BIGs).

FLAT ROOFS VERSUS PITCHED ROOFS

There is little doubt in terms of life expectancy, that a tiled/slate roof (pitched roof) is far superior to a flat roof. The most common faults with a flat roof are as follows:

(a) Making flat roofs totally without any drainage slopes was probably the main reason for there being so many leaks associated with flat roof construction in the past. The term 'flat roof' is inaccurate because a flat roof made from built up felt construction should never be flat.

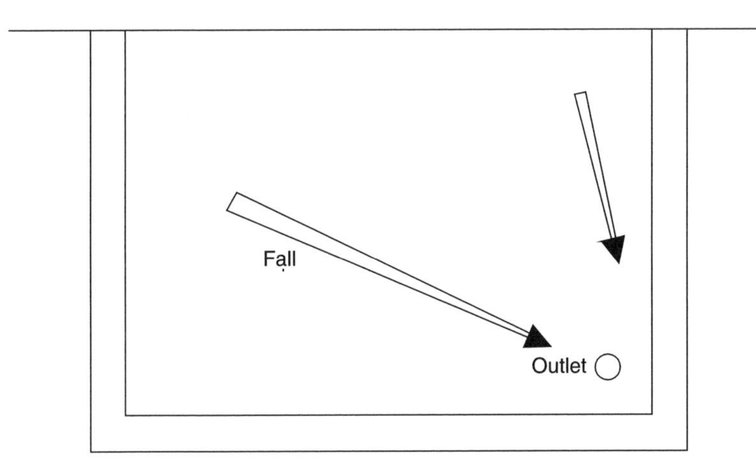

Fig. 18.2 Falls on flat roof directed to drainage point.

(b) If 'falls' were provided they were not large enough and water 'ponded' on the surface and eventually worked its way through the membrane. In order to drain satisfactorily the deck on domestic extensions should be laid to falls of minimum 1:40. One also has to remember that on large flat roofs it will be necessary to create 'falls' and 'crossfalls' (i.e. the roof water needs to be directed to a drainage point) (Fig. 18.2).

(c) The decking material used (chipboard or plywood) was the wrong grade of material for roofing and deteriorated. Chipboard should be of Grade C4. Plywood should be WBP grade. Chipboard should be supported at all edges by inserting additional noggins between the joists.

(d) The felt finish has not been laid in accordance with recommended standards.

(e) The surface of the felt has not been adequately protected from sunlight and as a result has flexed, bubbled and blistered.

The obvious answer to the criticisms mentioned above is to avoid flat roofs wherever possible.

19 Pitched roofs

GENERALLY

The term 'pitched roof' basically covers the standard tiled or slated roof that is the norm in the UK.

COVERINGS – TILES/SLATES (SEE FIGS 19.1 TO 19.5)

The coverings to a pitched roof comprise as follows

(a) A covering of manufactured tiles, shingles, or slates (natural or manufactured).
(b) Ridge and hip tiles to cover the exposed edges.
(c) Leadwork for flashings, soakers, aprons and the like.
(d) Battens to support the roof finish.
(e) Sarking or underfelt as a secondary barrier if a tile/slate is lost in a high wind.

The tiles or slates have to be laid to a suitable pitch (i.e. not less than the manufacturer's recommended instructions). If you refer to Figs 19.1 to 19.2, you will note that the minimum pitch is clearly indicated for each type of covering. The companies that produce roof coverings carefully test their products for weather resistance, and if their products are laid to slopes lower than the recommended minimum then there is no guarantee that they will provide the performance required.

Note: The recommended slope must be maintained over the whole roof pitch. It is normal practice to provide tilt fillets at the eaves in order that the tiles/slates discharge into the guuers properly. I have seen cases in my locality where the builders have forgotten this principle and have inserted an overlarge tilt fillet. This created a weakness at the change in slope and the houses affected by the design fault have suffered from water penetration at the eaves.

In general terms, the smaller the tile or slate, the steeper the pitch required. In order to avoid queries by Building Control, when specifying a roof finish, you must state the pitch on your plans.

Technical data

Size of tile	420mm x 330mm
Minimum pitch*	22.5° Smooth (75mm headlap) 17.5° Smooth (100mm headlap) 30° Granular (75mm headlap) 25° Granular (100mm headlap)
Maximum pitch	90°
Minimum headlap	75mm
Maximum gauge	345mm
Profile depth	d < 5mm
Cover width	292mm (nominal)
Hanging length	397mm (nominal)
Covering capacity (net)	9.9 tiles/m² at 75mm headlap 10.7 tiles/m² at 100mm headlap
Weight of tiling (approx.)	50kg/m² (0.49 kN/m²) at 75mm headlap 54kg/m² (0.53 kN/m²) at 100mm headlap
Battens required (net)	2.9 lin.m/m² at 75mm headlap 3.1 lin.m/m² at 100mm headlap
Batten size recommended (fixed to BS 5534)	38 x 25mm for rafters/supports not exceeding 450mm centres 50 x 25mm for rafters/supports not exceeding 600mm centres
Tile nails	50mm x 3.35mm
Fixing clips	Eaves, verge and tile clips

* The minimum recommended pitch and lap may be influenced by special circumstances, please contact the Technical Advisory Service.

Fig. 19.1 Example of technical data for Marley Modern Tiles. (Reproduced with the permission of Marley Tile Company.)

Slate size (nominal) mm	Inches	Moderate exposure: driving rain Index less than 7 m²/s Minimum rafter pitch									Severe exposure: driving rain Index 7 m²/s or more Minimum rafter pitch								
		20°	22½°	25°	27½°	30°	35°	40°	45°	85°	20°	22½°	25°	27½°	30°	35°	40°	45°	85°
660 × 355	**26 × 14**	130*	105	90	80	75	75	65	65	–		140	120	115	105	85	75	65	–
610 × 355	**24 × 14**	115	105	90	80	75	75	65	65	–	145*	125	105	95	90	75	75	65	–
610 × 305	**24 × 12**	130*	115	90	80	75	75	65	65	–	–	120	120	115	110	90	80	70	–
560 × 305	**22 × 12**	115	105	90	80	75	75	65	65	–	140*	120	105	100	90	75	75	65	–
560 × 280	**22 × 11**	120*	110	90	80	75	75	65	65	–	145*	130	110	105	100	85	75	65	–
510 × 305	**20 × 12**	115	105	90	80	75	75	65	65	–	–	125	100	95	85	75	75	65	–
510 × 255	**20 × 10**	125*	110	90	80	75	75	65	65	50	–	135	115	110	100	90	75	65	65
460 × 305	**18 × 12**	115*	105	90	80	75	75	65	65	50	–	–	110	95	85	75	75	65	65
460 × 255	**18 × 10**	125*	110	90	80	75	75	65	65	50	–	–	115*	110	100	85	75	65	65
460 × 230	**18 × 9**	125*	115*	100	80	75	75	65	65	50	–	–	120*	115*	105	95	85	65	65
405 × 305	**16 × 12**	–	–	–	80	75	75	65	65	50	–	–	–	–	90	80	75	65	65
405 × 255	**16 × 10**	–	–	–	85	75	75	65	65	50	–	–	–	–	100	95	75	65	65
405 × 230	**16 × 9**	–	–	–	85	75	75	65	65	50	–	–	–	–	100	100	85	65	65
405 × 205	**16 × 8**	–	–	–	90	75	75	65	65	50	–	–	–	–	105	100	90	65	65
355 × 305	**14 × 12**	–	–	–	80	75	75	65	65	50	–	–	–	–	75	75	75	65	65
355 × 255	**14 × 10**	–	–	–	80	75	75	65	65	50	–	–	–	–	80	75	75	65	65
355 × 230	**14 × 9**	–	–	–	80	75	75	65	65	50	–	–	–	–	85	80	75	65	65
335 × 205	**14 × 8**	–	–	–	80	75	75	65	65	50	–	–	–	–	90	85	75	65	65
355 × 180	**14 × 7**	–	–	–	80	75	75	65	65	50	–	–	–	–	95	90	80	65	65
305 × 255	**12 × 10**	–	–	–	80	75	75	65	65	50	–	–	–	–	75	75	75	65	65
305 × 205	**12 × 8**	–	–	–	80	75	75	65	65	50	–	–	–	–	85	80	75	65	65
305 × 150	**12 × 6**	–	–	–	80	75	75	65	65	50	–	–	–	–	85	80	75	65	65
255 × 255	**10 × 10**	–	–	–	80	75	75	65	65	50	–	–	–	–	75	75	75	65	65
255 × 205	**10 × 8**	–	–	–	80	75	75	65	65	50	–	–	–	–	75	75	75	65	65
255 × 150	**10 × 6**	–	–	–	80	75	75	65	65	50	–	–	–	–	75	75	75	65	65

Note: the actual pitch at which the slate lies on the roof is less than the rafter pitch by an amount which is a function of the slate thickness and the lap. Therefore head laps marked* are not suitable for use with extra heavy slates.

Fig. 19.2 Slate roofing dimension details. (Reproduced with permission of Alfred McAlphine Slate Products Ltd.)

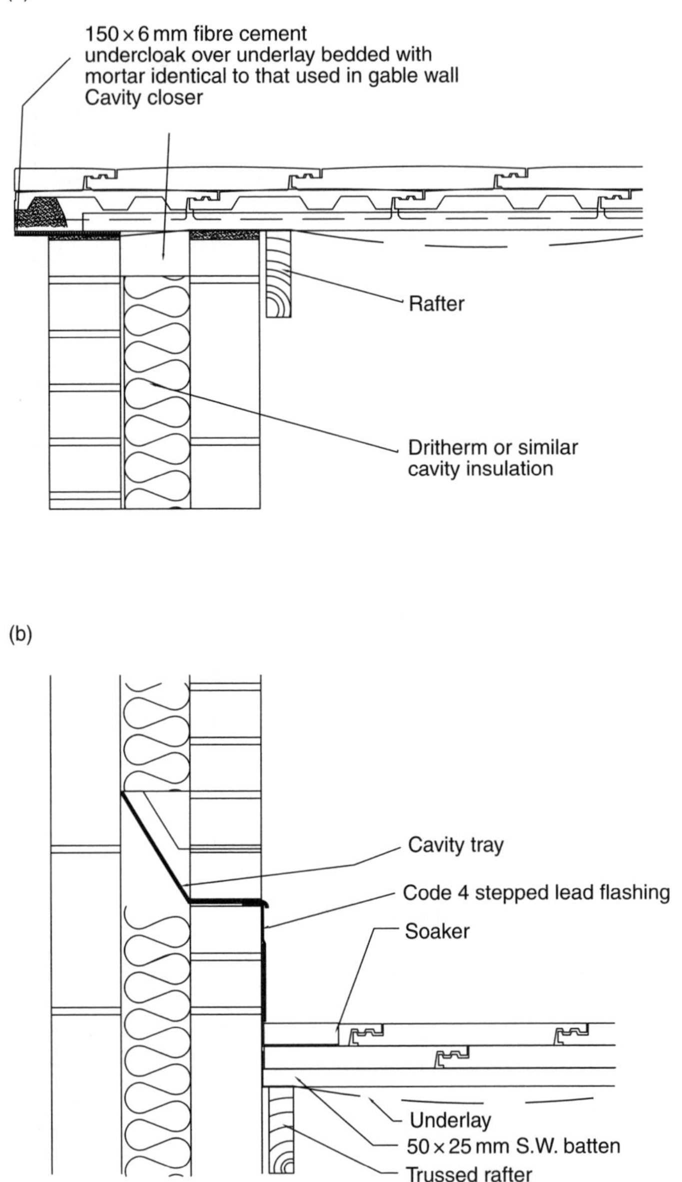

Fig. 19.3 Roof details: (a) mortar bedded verge; (b) side abutment; (c) typical GRP Valley trough; (d) typical ridge detail.

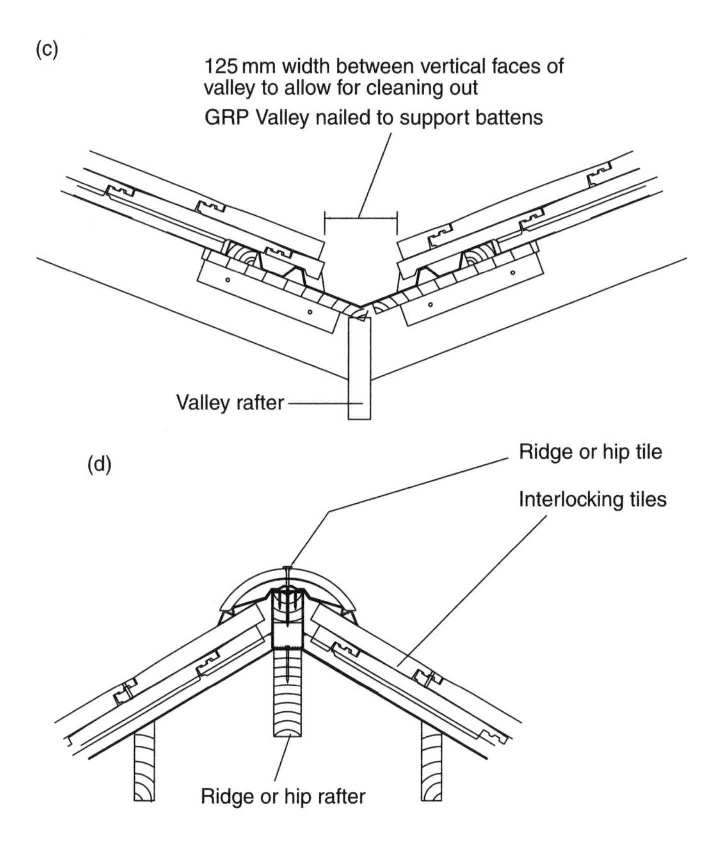

Fig. 19.3 contd.

As examples, I have extracted a list of a few tiles and lowest pitch as follows:

As there are many manufacturers and a variety of different roof tiles/slates on the market, it would be advisable for you to contact the various

Tile type	Mm. pitch (degrees)
Marley Wessex	15
Marley Modern, Smooth	17.5
Redland Stonewold, Delta Regent	17.5
Marley Modern, Granular	22.5
Redland Renown	30
Marley Plain	35
Redland Rosemary	35
Imitation slate type	
Marley Monarch	22.5
Bradstone Moordale	25
Bradstone Cotswold	30

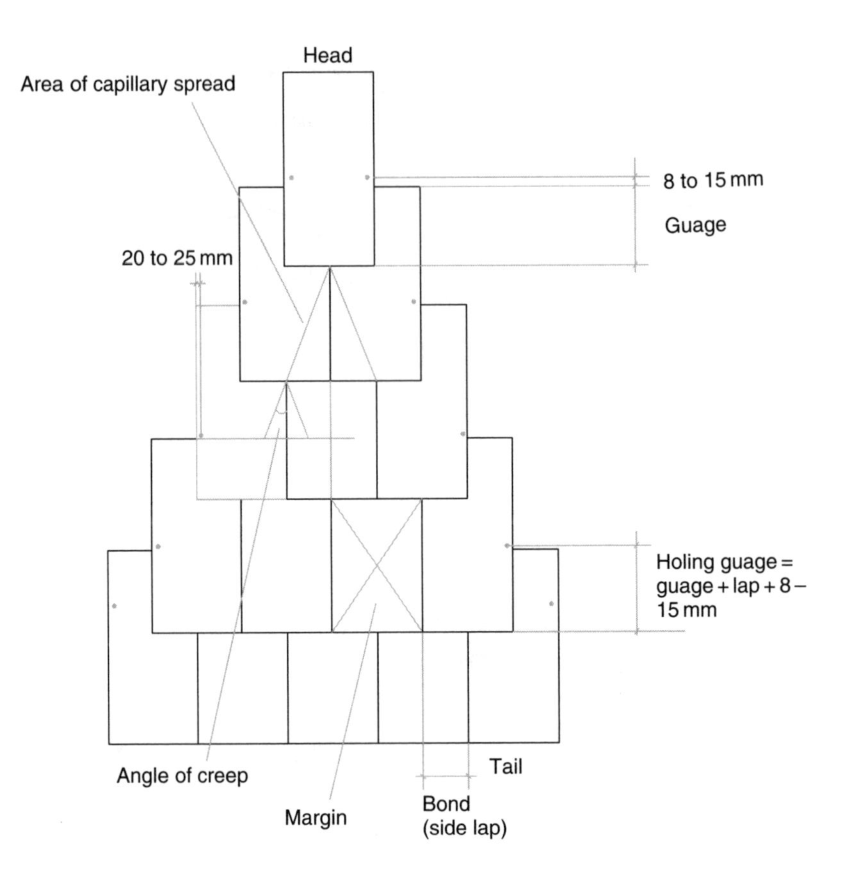

Fig. 19.4 Typical slate roof.

manufacturers and request copies of their current catalogues. These will give you a complete performance specification and descriptions to use on your plans for each type of tile/slate. This information includes such matters as minimum head and side laps. For examples see Fig. 19.1.

Natural slates

Natural slates, whether new or secondhand, come in a variety of sizes (Fig. 19.2), and roofs will leak if the slates are too small in relation to their pitch, or slates are not wide enough in relation to their length.

Tile/slate battens

The roof tiles or slates are fixed to battens over sarking felt. With regard to sizes of battens required to support the covering, the tile or slate manufacturer's recommendations should be followed.

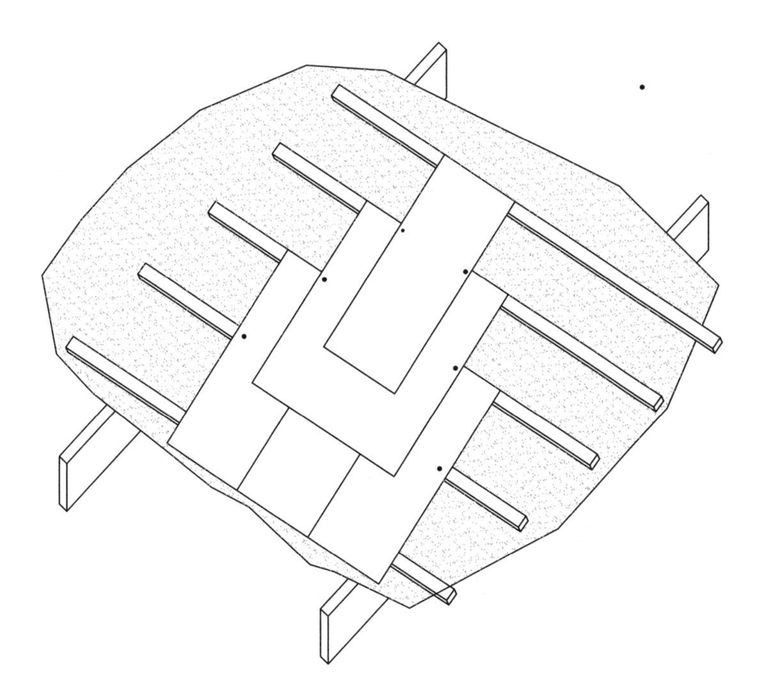

Fig. 19.5 Slate roof details.

PITCHED ROOF STRUCTURE

There are two basic types of pitched roofs now in common usage:

(a) traditional type
(b) trussed rafter type (Fig. 19.8).

I have also included a short section on SIPs because there is a possibility that SIPs may replace both (a) and (b) where roof spaces are used for habitable purposes (Fig. 19.9).

Traditional roof

Below the roof finish (i.e. the tiles/slates and underfelt) the traditional roof comprises a framework of timbers. Each individual member has its own traditional name (Figs 19.6 and 19.7). Sizes of roof members required within a roof can be obtained from TRADA tables.

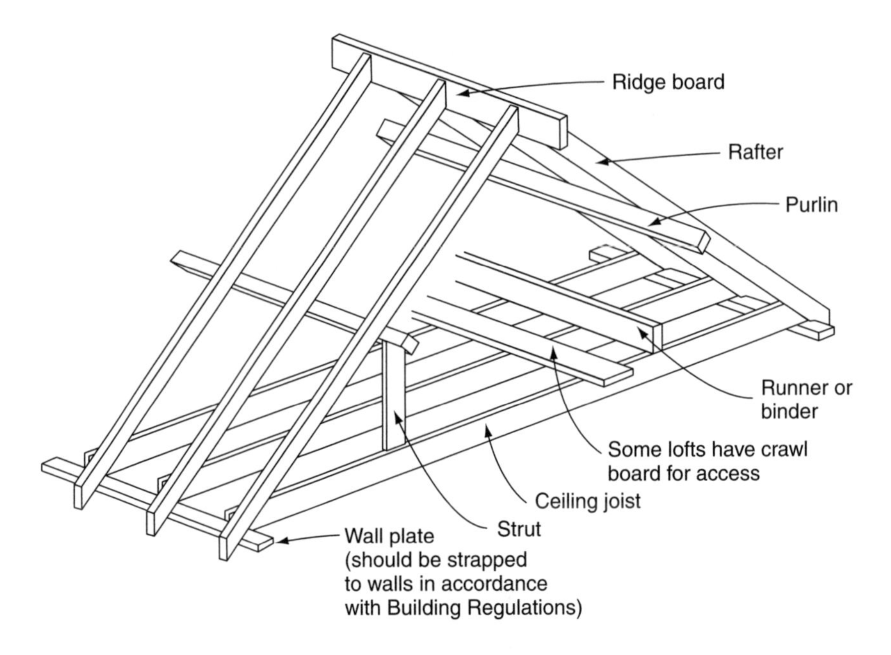

Fig. 19.6 Names of roof timbers in a traditional roof.

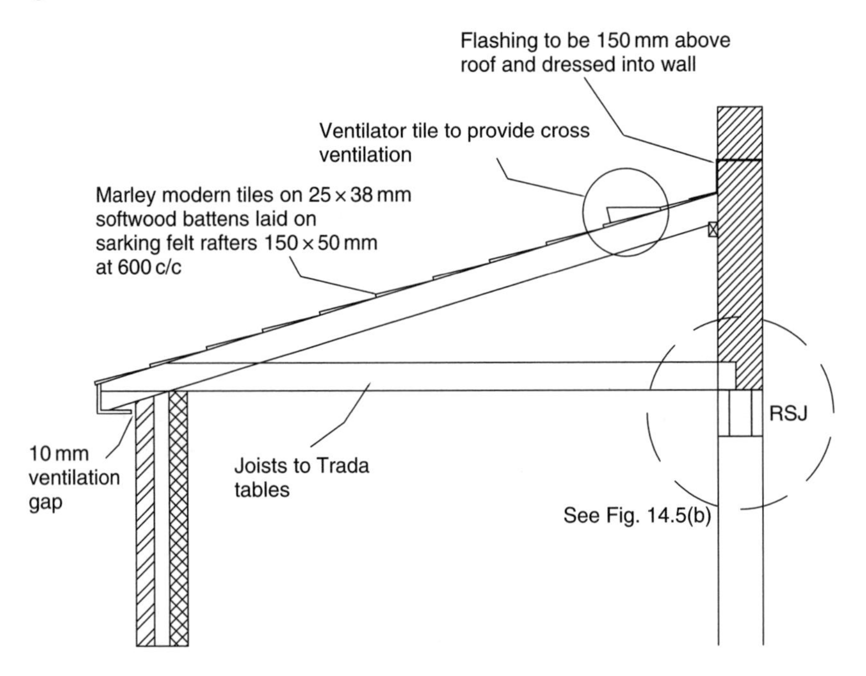

Fig. 19.7 Typical section through lean to extension showing pitched roof over 15 degrees.

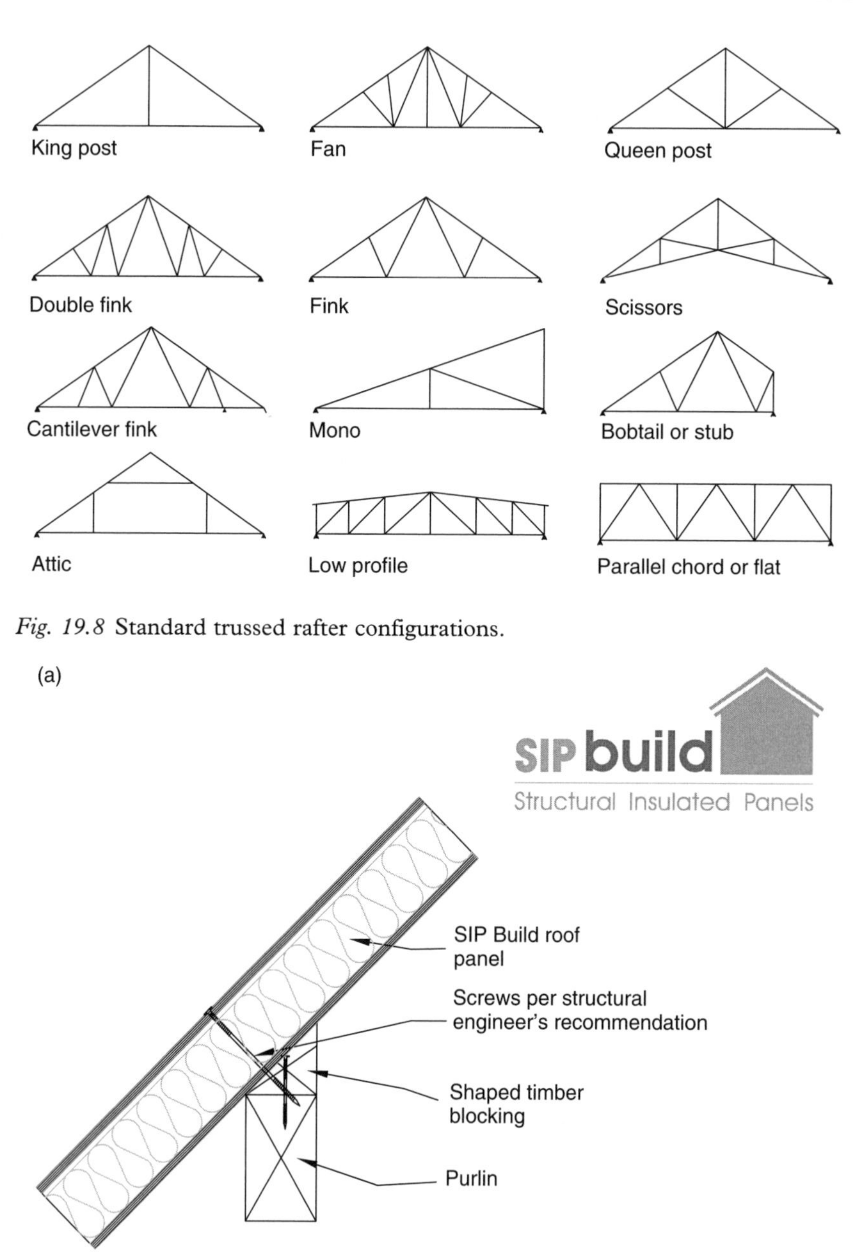

Fig. 19.8 Standard trussed rafter configurations.

Typical purlin connection detail

Fig. 19.9 SIP roof details: (a) typical purlin connection detail; (b) typical steel purlin connection detail; (c), (d) typical roof finish detail; (e) typical ridge detail; (f) typical eaves detail; (g) typical valley detail.

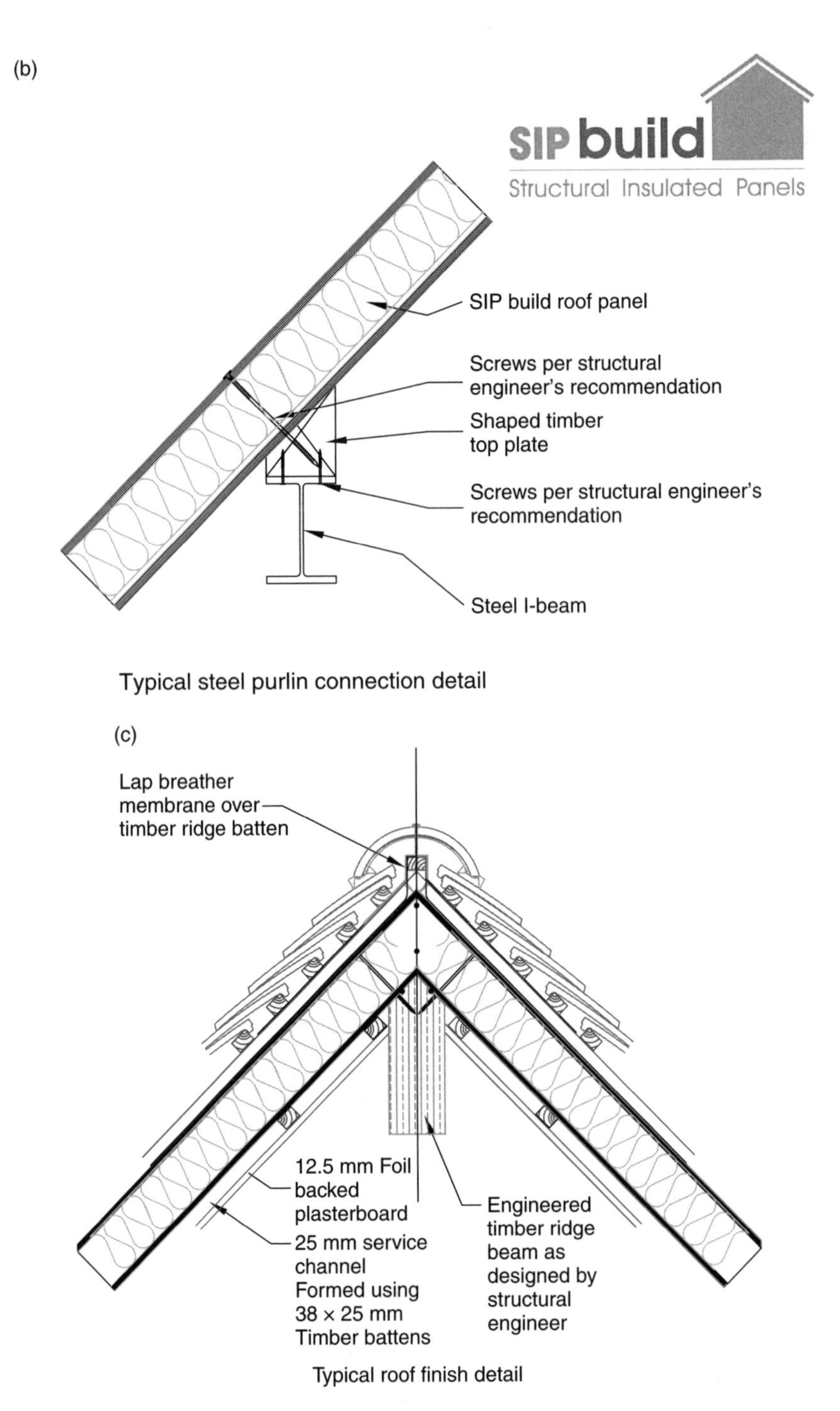

Fig. 19.9 contd.

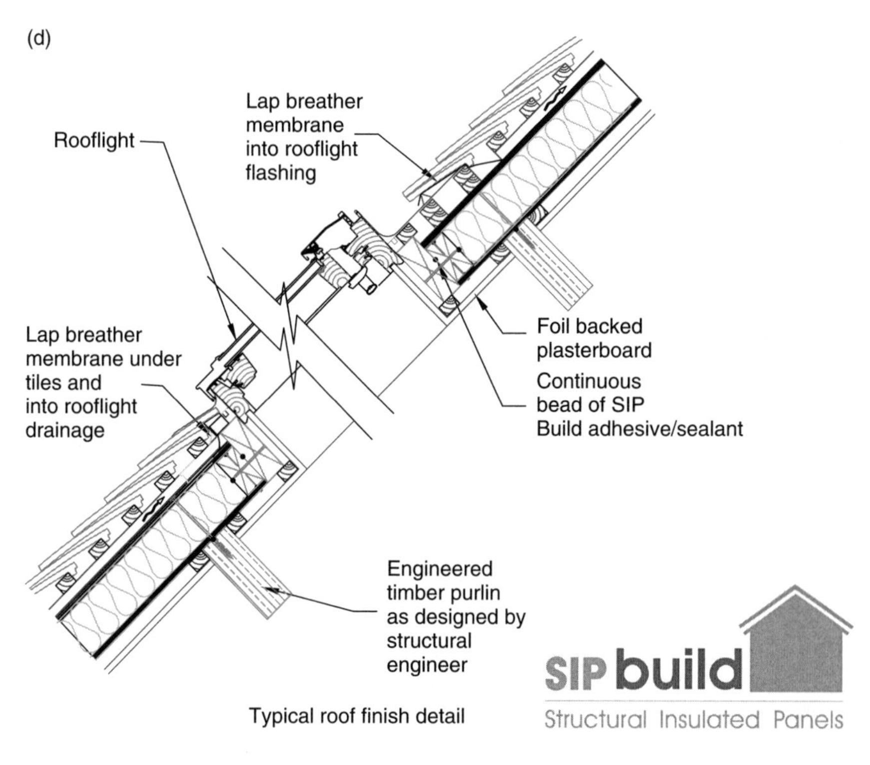

Fig. 19.9 contd.

Their common names and their functions are as follows:

(a) Rafters (common rafters) – these are usually spaced at 400 mm centres and carry the weight of the roof finish.
(b) Ceiling joists – for roofs of any reasonable span, the ceiling joists perform two functions

  (i) to hold up the ceiling;
  (ii) to act as ties and prevent the roof spreading by forming the third side of the triangle.

(c) Purlins – heavy timber beams that support the rafters. They are not always needed on smaller roofs.
(d) Hangers, binders and struts – subsidiary members that help to further triangulate the roof and strengthen it.
(e) Ridge board – this member gives stability to the rafters and also a fixing for the cut ends of common rafters.

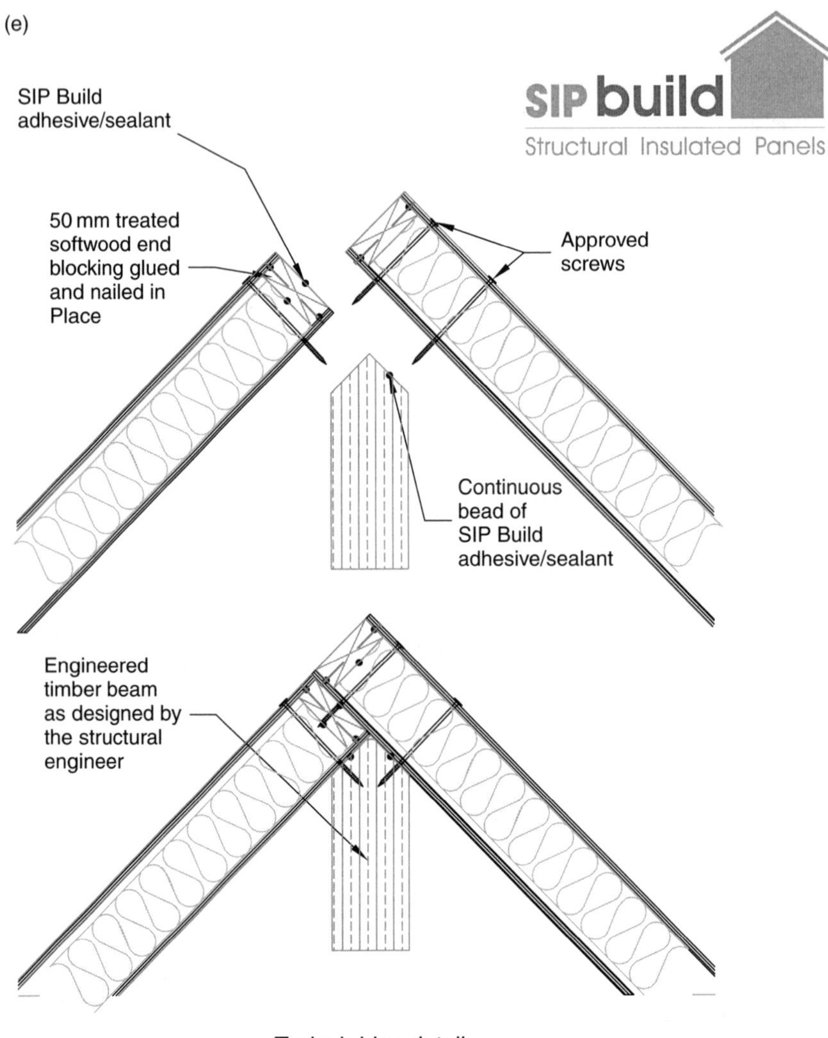

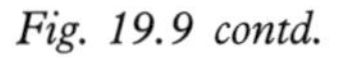
Fig. 19.9 contd.

(f) Hip rafter – a large rafter that runs underneath a hip on a hipped roof and carries the cut ends of common rafters.
(g) Valley rafter – a large rafter that runs underneath a valley on a roof and carries the cut ends of common rafters.
(h) Wall plate – a horizontal timber bedded onto the top of the brickwork to ensure that the top of the wall is level.

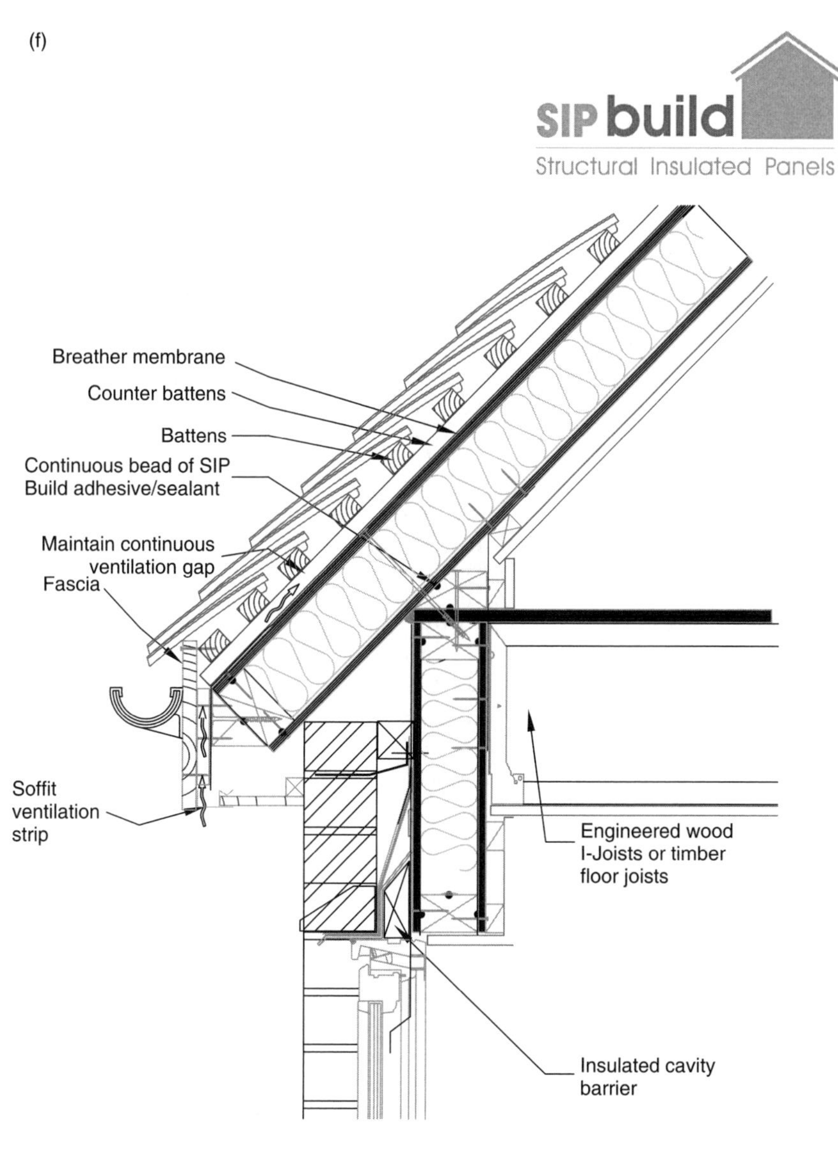

Typical eaves detail

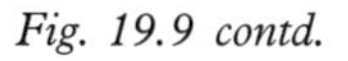

Fig. 19.9 contd.

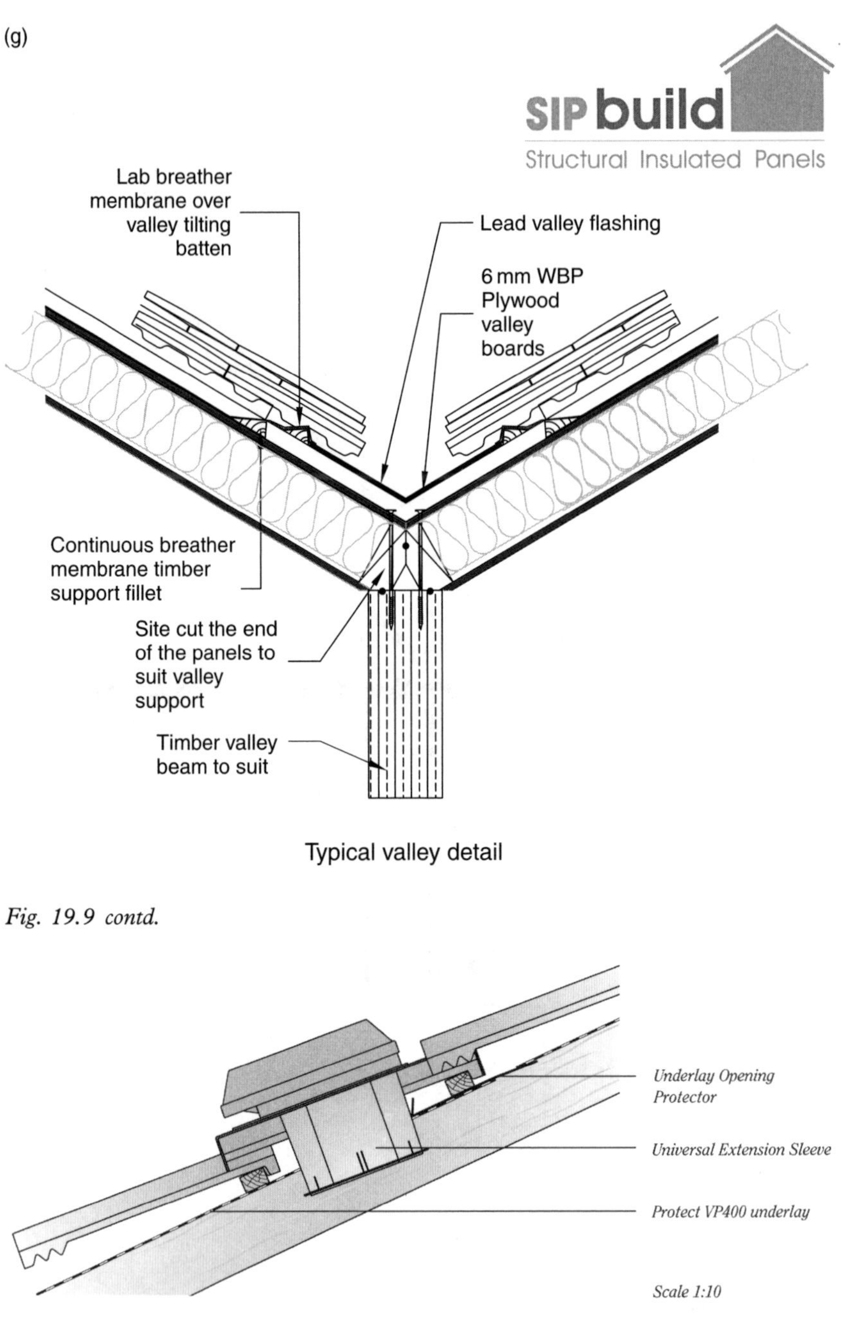

Fig. 19.9 contd.

Fig. 19.10 Typical roof ventilator. (Reproduced with permission of BDP-formerly William Building Products.)

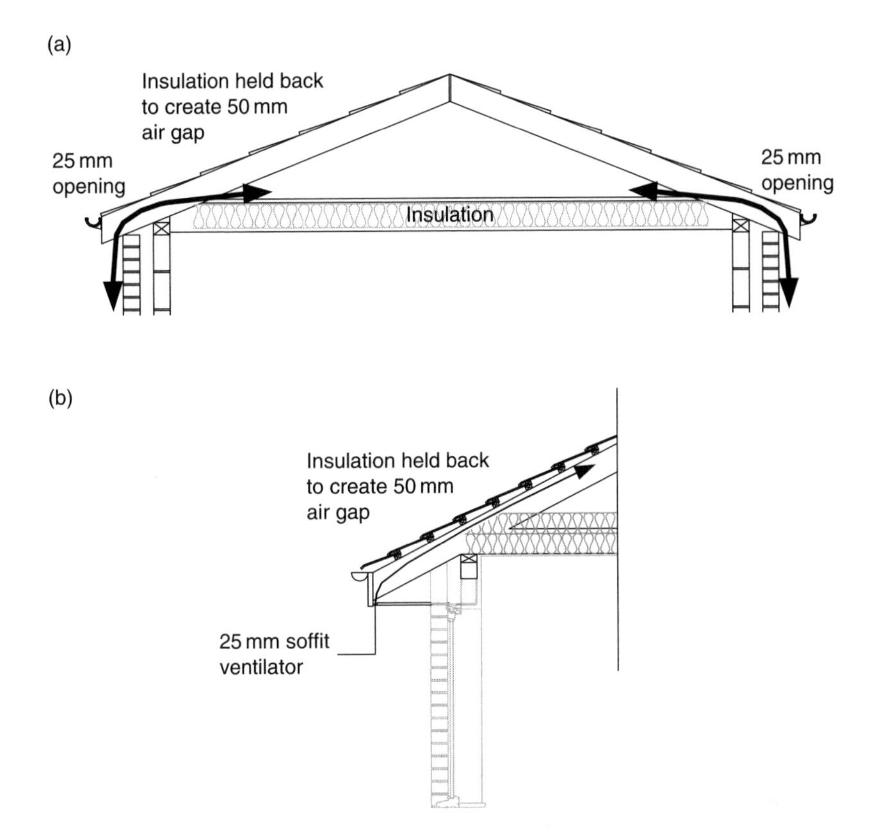

Fig. 19.11 Roof ventilation: (a) duopitch roof below 15 degrees; (b) detail of eaves.

Roof ventilation tile/slate – see Figs 19.10 to 19.12.
Note: One of the most glaring defects of the last few editions of the Building Regulations is that they ignore elements (f) and (g) above. The NHBC also appear to have dodged the issue and have placed roof design, with regard to hips and valleys, into the hands of an engineer. However, Appendix 7.2 of NHBC Standards is of some help, even though it does not provide the complete answer. The NHBC's minimum requirements are as follows:

- Struts and braces – 100 mm × 50 mm.
- Hips – minimum size should be the width of the cut end of the rafter plus 25 mm (no width is given).
- Valleys – minimum size should be 32 mm thick (no depth is given).
- Ridge board – minimum size should be width rafter cut plus 25 mm (no depth is given).

In the past, when queries have been raised by various Building Control Departments concerning sizes of these various elements, and I have tried

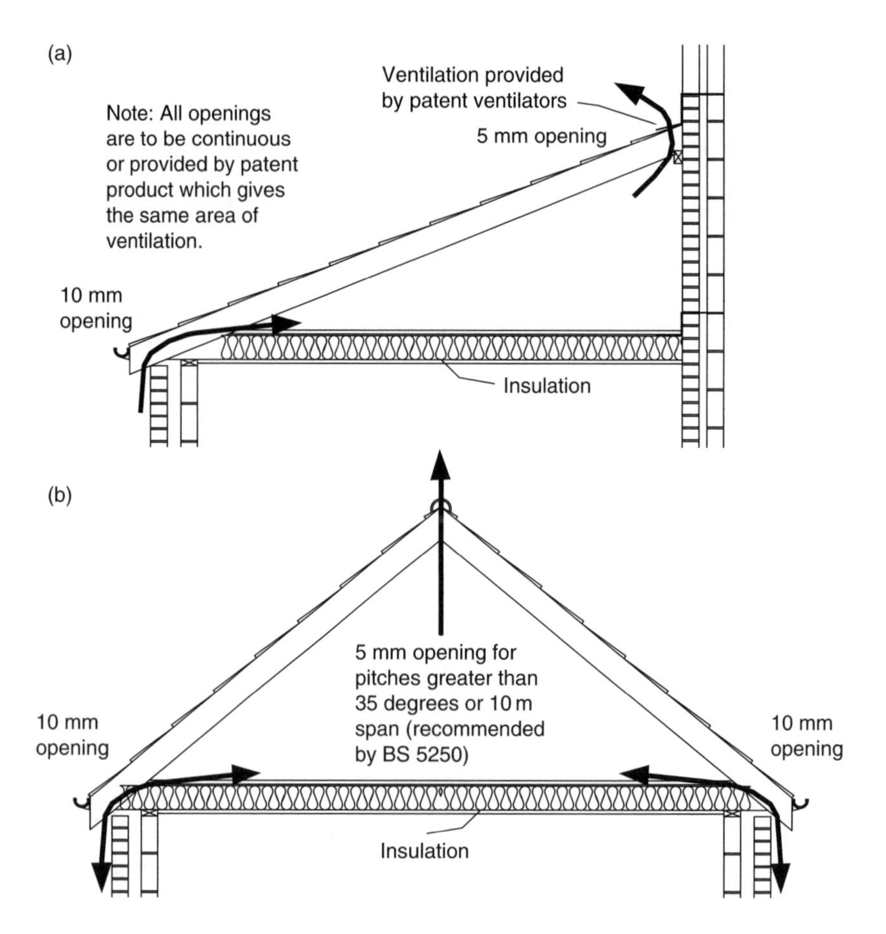

Fig. 19.12 Roof ventilation: (a) monopitch roof over 15 degrees; (b) duopitch roof over 15 degrees.

to substantiate the sizes that I have given, the only simple guidelines that anyone seems to be able to unearth is information contained within a very outdated copy of some old building byelaws (believed to be circa 1936) which gives sizes on roofs of 35° and 45° pitches which are as follows:

Valley rafters

(a) Make 1.50 in (40 mm) thick and twice the depth of common or jack rafter where the valley is supported intermediately by purlins which latter are strutted under junction with valley rafter, or
(b) Make 2 in (50 mm) thick and same depth as specified for ridges and hip rafters where valley rafter is intermediately supported as (a).

(**Note**: Valley rafters are beams subject to considerable loads and adequate strength is essential.)

Hip rafters and ridges (dimensions in inches (mm))

Depth of rafters	35° roof pitch	45° roof pitch
4 (100)	7 × 1.25(175 × 32)	8 × 1.25(200 × 32)
5 (125)	8 × 1.25(200 × 32)	9 × 1.25(225 × 32)
6 (150)	9 × 1.25(225 × 32)	11 × 1.25(300 × 32)

Note: The figures in brackets are my approximate metric conversions. It should be borne in mind that some BCOs may not be prepared to accept the use of old tables. I normally indicate that wall plates should be 100 mm × 50 mm.

TRUSSED RAFTER ROOF

With trussed roofs, the rafters and other elements used in a traditional roof are substituted by factory designed and manufactured roof trusses. In order that a trussed rafter roof can be passed by Building Control, you must obtain calculations from the truss manufacturers to support your application. As most truss manufacturers will not provide calculations until they receive a firm order, you can ask Building Control to approve the plans on the understanding that the calculations will be provided before the roof work starts on site (a conditional approval).

Trussed rafters cannot be made on site (unless the contractor has the necessary plant – which is very unlikely). Trusses are factory-made components, manufactured to BS 5268. Trussed rafters are supplied in a variety of types (Fig. 19.8 gives the names of the most common trussed rafter types). The structure of a trussed rafter roof comprises the following elements:

(a) The roof trusses are usually spaced at 600 mm centers and come in a variety of configurations (see Fig. 19.8).
(b) Timber bracing (minimum 100 mm × 25 mm cross section) stabilizes the roof trusses. This bracing must not be installed in a haphazard fashion. Appendix 7.2C of NHBC Standards indicates the requirements for various roof layouts.

SIPs – USING STRUCTURALLY INSULATED PANELS IN THE ROOF OF AN EXTENSION

The use of SIPs has already been briefly discussed in Chapter 14. To recap, Structurally Insulated Panels consist of two outer boards of orientated strand board (OSB) sandwiching an inner core of Expanded Polyurethane MCFC Free, ODP ZERO. According to SIPBuild, the core material has a ozone depletion rating of zero and the insulation benefits of the panel 150 mm thick is 0.17w/mk, which is far better insulation standard than that currently required by the Building Regulations.

Creating a basic roof structure is simple using SIPs (see Fig. 19.9). As SIPs are extremely strong, and can easily shrug off normal loadings such as wind, snow and the weight of standard roof tiles or slates, they will provide a 4 m clear span in most domestic circumstances. This means that unlike a traditional roof, there is no need to provide rafters/trussed rafters to support the roof. Any spans over 4 m can be accommodated by inserting purlins. Even better, as the SIPs contain insulating material, there is no need to provide another form of insulant as long as the correct grade of SIPs is used in the construction. There is of course another advantage to using SIPs in a roof. As there are large numbers of people wanting their rooms in the roof to follow the roof line and give them more height, this happens automatically with SIPs. The writer would like to thank Peter Barr of SIPBuild for his help, technical expertise and assistance in writing this chapter.

INSULATION (GLASS FIBRE) AND VENTILATION

If glass fibre is used to insulate a modern roof, then to comply with modern regulations it will have to be between 270 mm to 300 mm thick. A typical specification would be two layers of crown Loft roll 40, 100 mm between ceiling joists and 200 mm laid over the joists.

20 Finishes etc.

GENERALLY

When dealing with small plans, it is unusual to become involved with internal finishes or layouts and I deal with them in one composite description. Refer to standard specification clause 33 in Appendix B.

As indicated in previous chapters where upper floors require sound insulation, the plasterboard ceiling has to have a mass of 10 kg/m^2. One way of complying is to line the ceiling with Gyproc Wallboard 10 (or similar) which is designed to comply with the regulations.

In other locations, ceilings with joists up to 450 mm centres – use 9.5 mm plasterboard. With roof trusses at 600 mm centres – specify 12.5 mm plasterboard.

Useful Weblinks

http://www.knaufinsulation.co.uk/output/design/page_165.html
http://www.british-gypsum.bpb.com/products/plasterboard_accessories/gyproc_acoustic/gyproc_wallboard_ten.aspx

21 Ventilation

APPROVED DOCUMENT F

Approved Document F (2006 edition) is substantially different than its predecessors. Now it only deals with ventilation for people within a building. Roof ventilation is covered within Approved Document C.

Part F requires a building to have:

(1) extract ventilation
(2) whole building ventilation
(3) purge ventilation.

Diagram 1 on page 12 of the Regulations shows diagrammatically various acceptable ventilation systems for dwellings without basements. It is now possible to install:

(1) System 1 – Background ventilators and intermittent extract fans
(2) System 2 – Passive stack ventilation
(3) System 3 – Continuous mechanical extract
(4) System 4 – Continuous mechanical supply and extract with heat recovery.

As this book deals with extensions to existing properties I shall opt for System 1. Please note, I have presumed that non-habitable rooms will have windows and have indicated that they will have opening casements because most normal householders would have these to be provided. My simplified interpretation of the regulations is as follows:

Ventilation to rooms

Room	Purge (rapid) ventilation (e.g. opening windows or doors) percentage assumes window/door opens more than 30 degrees	Background ventilation	Intermittent extract ventilation
Habitable room	5% of floor area	5000 mm^2	
Kitchen	5% of floor area	2500 mm^2	30 L/S to hob or 60 LS if elsewhere

Contd.

Utility room	5% of floor area	2500 mm²	30 L/S
Bathroom	5% of floor area	2500 mm²	15 L/S
Sanitary accommodation (separate from bathroom)	5% of floor area Mechanical extract of 6 L/S	2500 mm²	

Addition of a habitable room to an existing building

Where one room obtains its ventilation via another room (it has no windows of its own), the opening between them must be at least 5 per cent of the floor area of both rooms combined and the room with the window or door in it also has an opening equal in area to 5 per cent of the floor area both rooms combined, and trickle ventilation not less than 8000 mm^2 (see Diagram 2 on page 21 of Approved Document F).

Habitable rooms can be ventilated via a conservatory as long as there are ventilation openings in both the conservatory and the habitable room (e.g. Patio doors) which have an opening area equal to 5 per cent of the floor area of the conservatory and room added together, and trickle ventilation not less than 8000 mm^2 in both the conservatory and the habitable room (see Diagram 3 on page 21 of Approved Document F).

Replacement windows

Clauses 3.3 to 3.6 cover the replacement of existing windows. Where a window is replaced, it should comply with the requirements of Approved Document L1B (Conservation of Fuel and Power) and Approved Document N (Glazing) and have trickle ventilation if needed. It is also a requirement that the ventilated opening shall be the same size as the original. If for some reason the original ventilation opening is not known (sometimes people remove casements and put in fixed lights instead) then the flowing sizes apply:

Habitable rooms	5000 mm²	E.g. 50 mm × 100 mm
Kitchen, utility room and bathroom with or without WC	2500 mm²	E.g. 50 mm × 50 mm

22 Staircases

REQUIREMENTS OF THE APPROVED DOCUMENTS

Staircases are covered by Approved Document K. I have not included details concerning winding stairs, spiral stairs or tread stairs (for details on these constructions refer to the document). On normal staircases Approved Document K dictates the following requirements (Fig. 22.1):

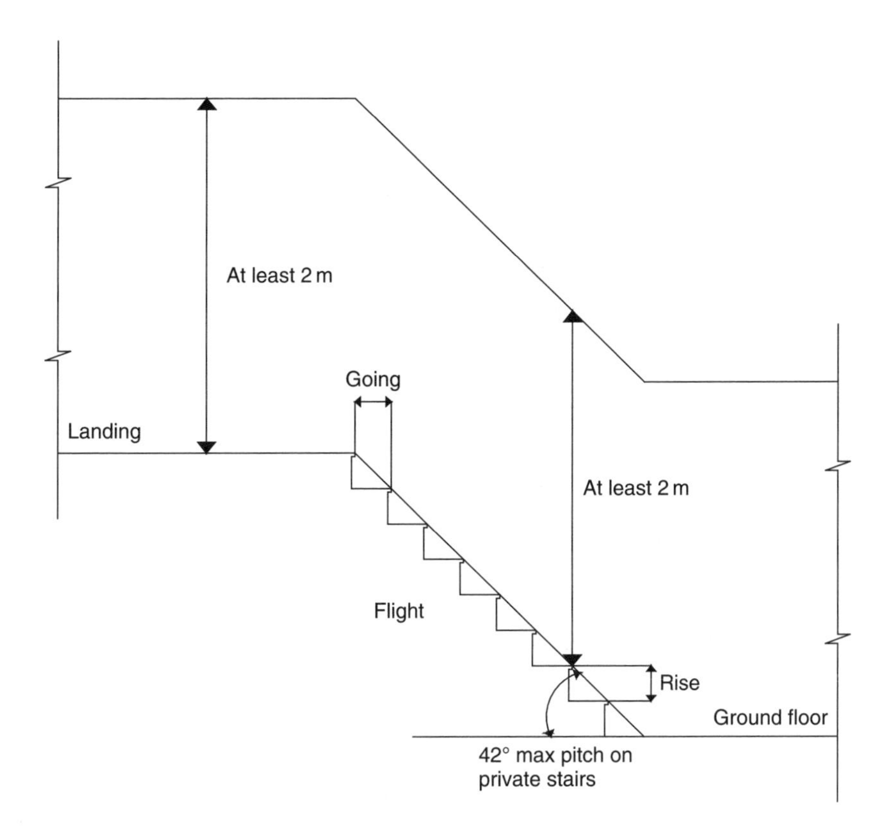

Fig. 22.1 Typical private staircase.

(a) Private stairs (stairs in houses) must not exceed 42° pitch.
(b) Stairs must have a minimum headroom of 2 m. (Slightly less headroom is allowed in loft conversions; see Diagram 3 of Approved Document K for details.)
(c) Twice the rise plus once the going should add up to 550–700 mm (maximum rise 220 mm, minimum going 220 mm).

  (i) Any rise between 155 mm and 220 mm used with any going between 245 and 260 mm will comply.
  (ii) Any rise between 165 mm and 200 mm used with any going between 223 mm and 300 mm will comply.

(d) No width given (800–900 mm is normal).
(e) Where a stair has an open side (not between walls) and the drop to the floor is more than 600 mm (or the stair is over 1000 mm wide) a

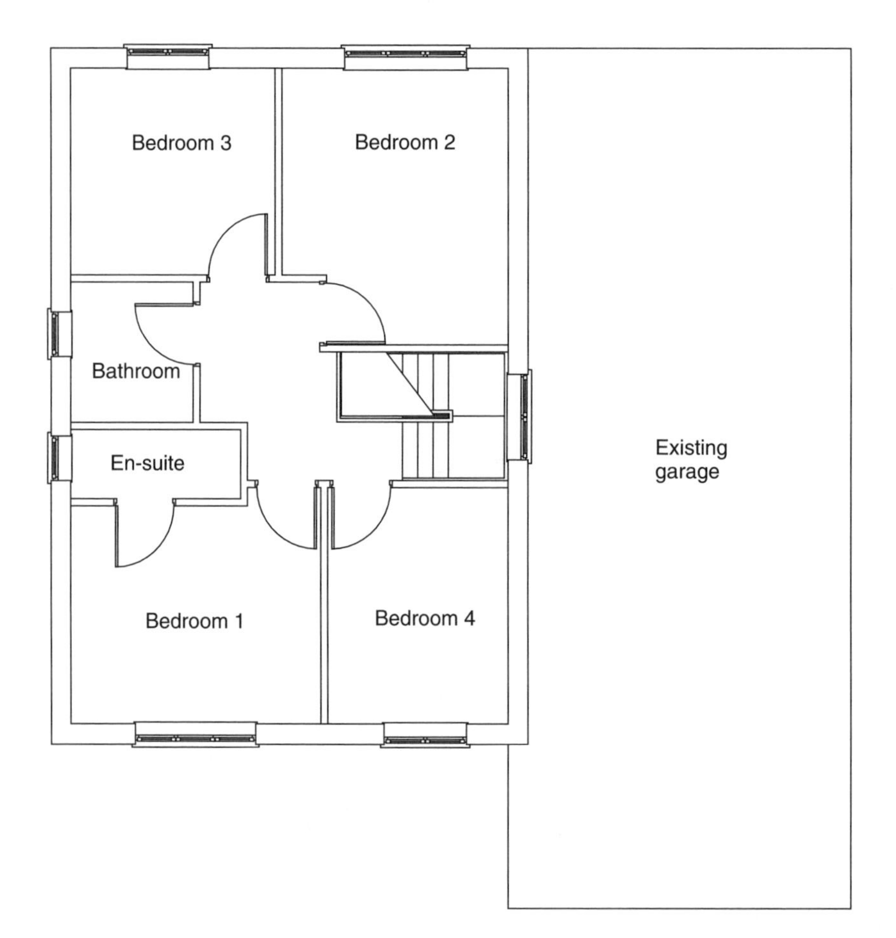

Fig. 22.2 Existing first floor plan.

handrail has to be provided. The handrail has to be between 900 mm and 1000 mm high. Upstairs landings should have the same protection.

(f) The balusters to a stair should not allow a sphere of 100 mm diameter to pass through.

(g) Landings must be provided top and bottom of stair flights. The width and length of each landing to be the same as the stair width.

When you are dealing with extensions to existing multistorey properties, it is unusual to become involved in installing a complete new staircase because the house already has one. A very common scenario, however, is the provision of short flights off an existing half landing.

If you refer to Figs 22.2 and 22.3, I have detailed a typical situation. Figure 22.2 shows the existing first floor plan and Fig. 22.3 shows the proposed first floor plan which indicates that the house will eventually have

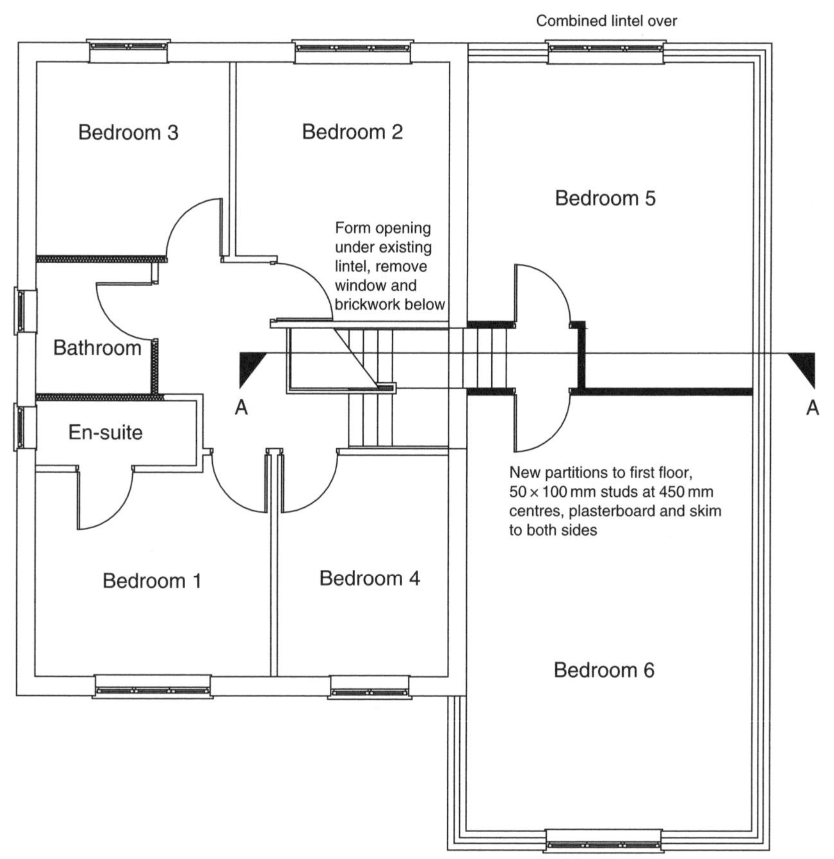

Fig. 22.3 Proposed first floor plan.

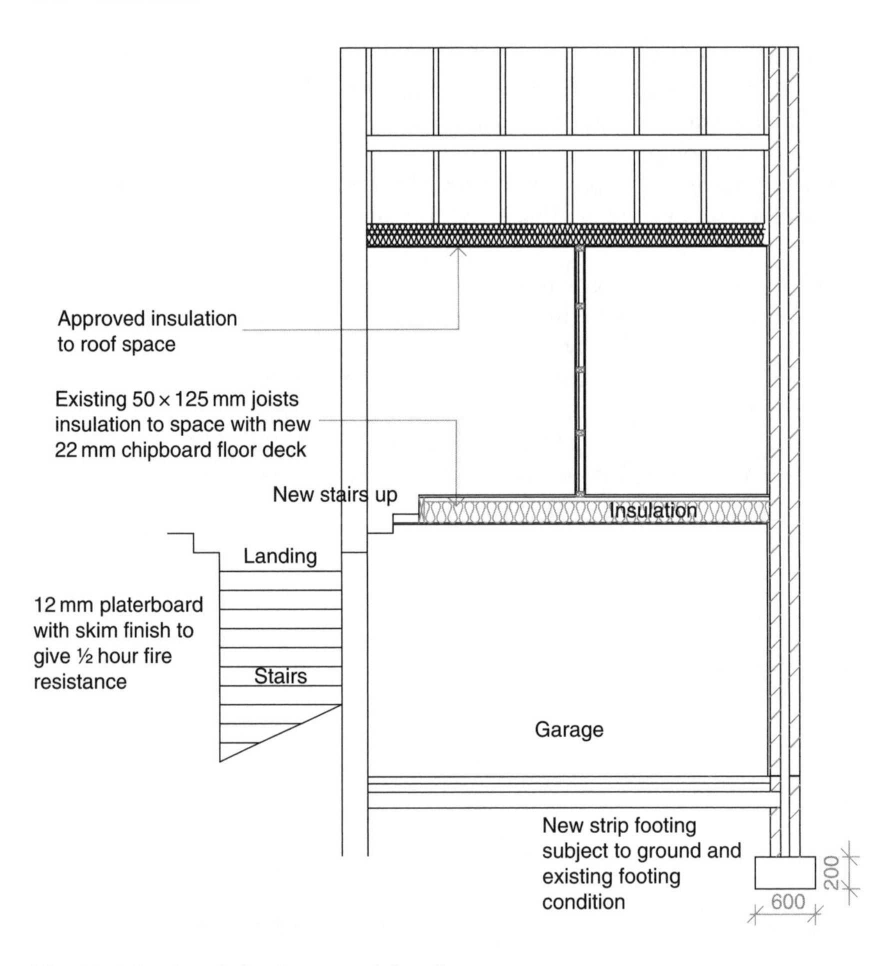

Fig. 22.4 Section A-A of proposed first floor.

two additional bedrooms. The original staircase is not a straight flight but stops at a half landing and then turns and rises again in a short flight until it meets the main landing. As the existing half landing is not at first floor height, there is only one economic solution (unless the existing bathroom or front bedroom is sacrifed) and that is to create a mirror image of the short flight of stairs and create a connection through what was the outer wall. Figure 22.4, Section A-A, indicates the proposal in more detail.

23 Plumbing, drainage, heating, installation of boilers, electrical works

ABOVE-GROUND DRAINAGE GENERALLY

Figure 23.1 shows a typical rear view of a house with the external pipework on view. This figure shows a typical 'old fashioned' system which has both a hopperhead to take the washbasin discharge and a separate soil and ventilation pipe (SVP) to take the toilet (WC) discharge. Nowadays, hopperheads should not be used and all the waste should be connected to the SVP. As rainwater drainage has already been covered in other sections, the only major item of above ground level drainage is the foul system (i.e. drainage from the WC and the like).

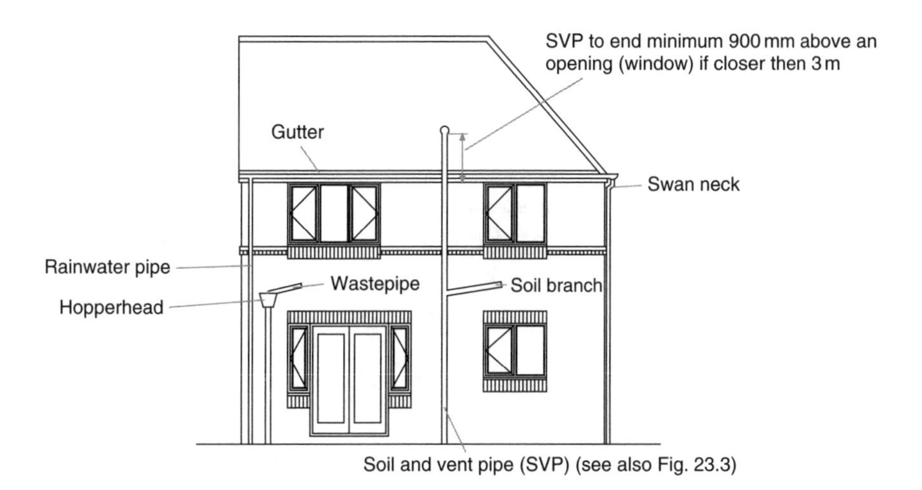

Fig. 23.1 External plumbing (old-fashioned system): hopperheads taking first floor waste should no longer be used.

TRAPS

As you will probably be aware, most sanitary fittings and sinks have a 'trap' underneath (a 'U' bend that holds water). Sanitary appliances and external gullies must have traps provided. The reason why they are there is simple: if they were omitted, foul air from the underground drains would enter the house, which is undesirable. Traps on pipes can be breached, however.

As anyone knows, who has studied physics at school or has ever made wine at home, if you want to get liquid out of one vessel into another, it is possible to make a simple syphon with a length of tube. Once the siphonage has started the liquid will empty into the second vessel without any problem as the weight of the water going down the tube (and the effects of atmospheric pressure) forces the liquid out of the first vessel. The principle of siphonage caused a problem for plumbers because sometimes on a long run of pipe, siphonage would start and empty every trap in the house, defeating their purpose (Fig. 23.2). In this sketch a householder has indulged in a little DIY work. Unfortunately, when the water starts to run down the overlong pipe, it is highly likely to syphon off all the water in the trap. In order to overcome this problem, there are restrictions on the lengths of pipes (Fig. 23.3).

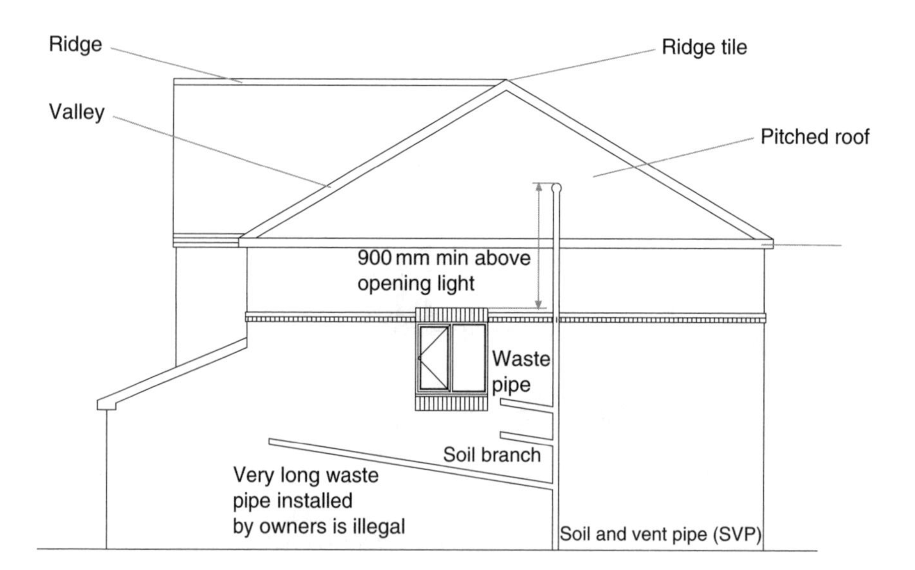

Fig. 23.2 External plumbing (single-stack system): overlong wastepipe causes siphonage problems.

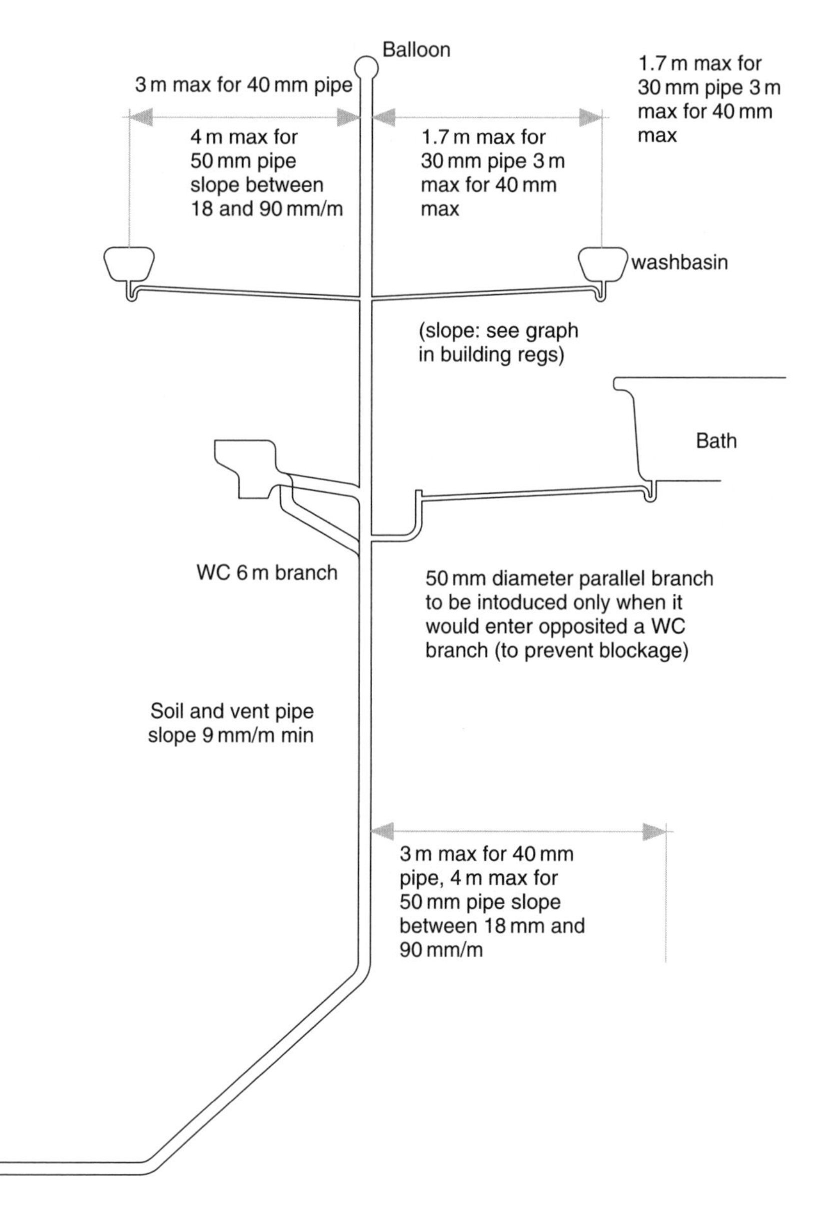

Fig. 23.3 Outline details of single-stack system showing length restrictions on pipes.

WASTE PIPES AND SOIL AND VENTILATING PIPES

It is a requirement of the Approved Document H that below-ground drains are ventilated to the open air so that explosive gases do not build up. A typical SVP system is shown in Figs 23.2 and 23.3.

Most buildings have a combined SVP because it 'kills two birds with one stone'. The SVP has an open end at high level which will discharge the foul air (the open end should be protected by a 'balloon' grating to stop birds nesting there) and at the same time allows wastes from WCs and wash basins, baths and the like to descend to the below-ground drains. Note that the upper end of all SVP must discharge to the open air in such a way that the foul smells do not come back into the house (i.e. if closer than 3 m to a window, it must extend 900 mm higher than the window).

Practical tip

Sometimes a new extension will be built so that the old SVP which was originally outside the building now becomes an internal pipe (Fig. 23.4). Don't just show the pipe boxed in. Indicate that a new plastic patent system will be installed (e.g. Key Terrain, Hunter Plastic etc.) because the old Cast-iron type SVP could leak and cause damage at some future date.

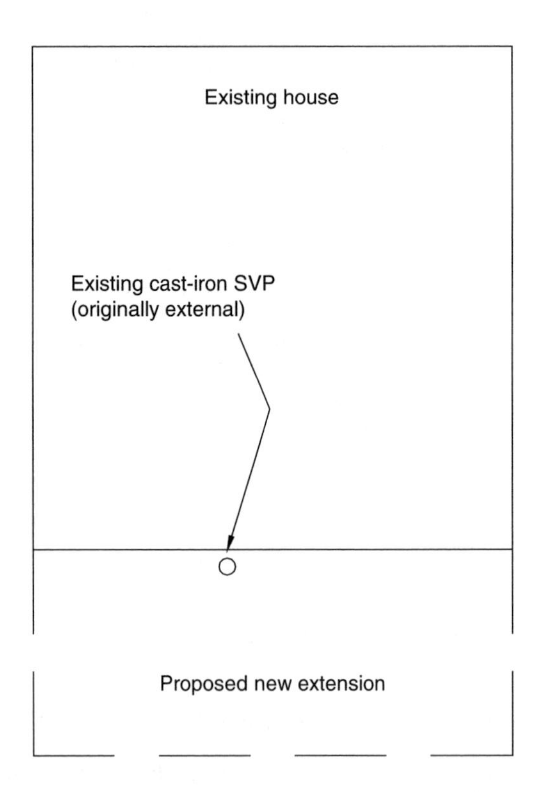

Fig. 23.4 External cast-iron SVP to existing house enclosed by new extension – replace with patent plastic system.

SPECIFIC REQUIREMENTS

Above ground

Table 2 of Approved Document H sets out the requirements for traps and pipework which are indicated on Fig. 23.3. If a sketch similar to this one is reproduced on drawings it indicates your intentions to the BCO and the builder.

Below ground

See Appendix B Specification clauses 45, 46 and 47.

ELECTRICAL WORKS – APPROVED DOCUMENT PART P

Generally

Electrical works are now covered by the Building Regulations (see Appendix B Specification clauses 35 to 40).

The requirements of Part P

The new regulations require

(1) Reasonable provision to be made in the design, installation, inspection and testing of electrical installations to protect persons against fire and injury.
(2) Where an electrical installation is provided, extended or altered, sufficient information to be provided so that persons wishing to operate, maintain or alter the installation in the future can do so reasonably safely.
(3) Part P only applies in England and Wales to fixed electrical installations after the electricity supply meter in all dwellings and

   (a) Dwellings and business premises that have a common supply;
   (b) Common access areas in blocks of flats;
   (c) Shared amenities of blocks of flats such as laundries and gymnasiums;
   (d) Outbuildings, including sheds, garages and greenhouses, and in garden areas (e.g. to ponds and lighting, supplied from a consumer unit located in any of the above).

Boilers

See Appendix B Specification clauses 41 to 44. All new and replacement natural gas and LPG boilers are required to have a minimum SEDBUK

(Seasonal Efficiency of Domestic Boilers in the UK) rating of 86 per cent. Oil fired boilers must have a minimum SEDBUK rating of 85 per cent. In order to achieve this, most boilers will have to be condensing boilers. There is also a requirement that new hot water radiators should be fitted with thermostatic radiator valves or be controlled by room thermostats.

24 Other building control aspects

BUILDING A SECOND-STOREY EXTENSION OVER AN EXISTING EXTENSION

Where a new second-storey extension is being built over an older extension, always assuming that the old wall construction is proved to be adequate (e.g. a suitable cavity wall), the Building Control Department will wish to ensure that the existing structure can take the new loads. With this in mind, they will often insist on a note on the plans requiring parts of the old footings to be exposed to prove adequacy.

OLD FOUNDATIONS

Prior to building control being applied to construction work, like most things, foundations were left to the knowledge and skill (or lack of) of the builder/ supervising officer. In Figs 24.1 and 24.2, I have shown typical older types of foundations and the adjacent floor construction. When

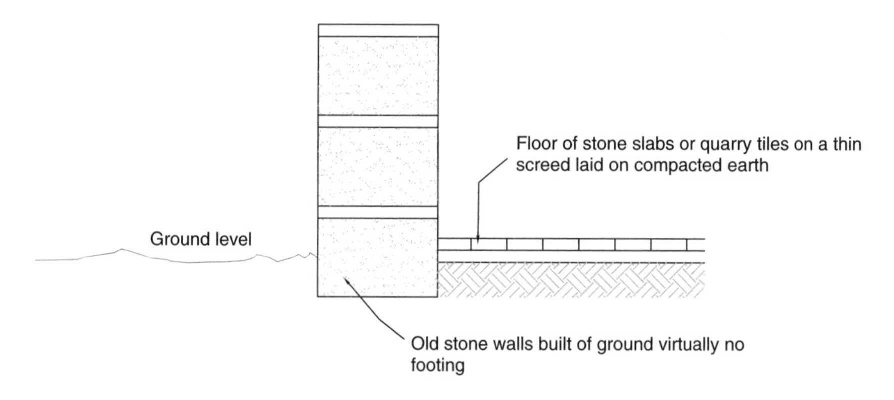

Fig. 24.1 Typical foundation in old cottage.

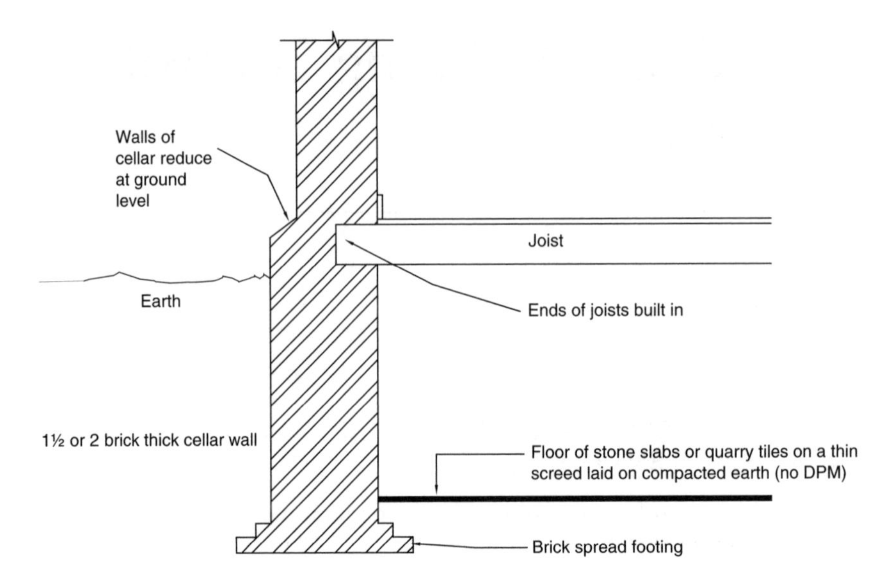

Fig. 24.2 Typical old cellar detail.

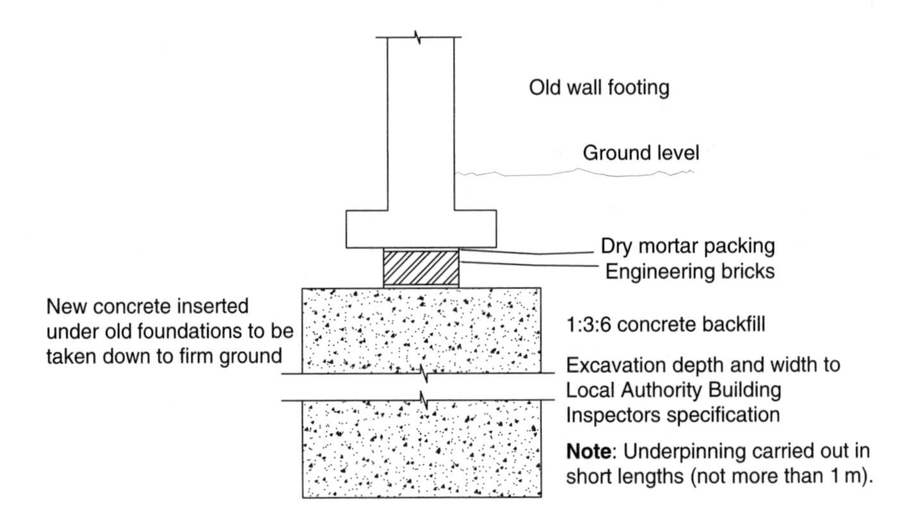

Fig. 24.3 Outline of underpinning detail.

building near old structures, especially ones that are showing signs of distress, it might be necessary to underpin some foundations (Fig. 24.3). (**Note**: The detail shown on Fig. 24.3 is only an outline sketch. If underpinning work is needed, it may be necessary to consult an engineer.)

THE SLOPING SITE

Sometimes people wish to build an extension or dwelling on a steeply sloping site. When this occurs, there could be problems with the foundations. Approved Document A indicates that the height between the floor slab on one side and the lower ground on the other should not exceed four times the wall thickness. The obvious solution to the problem is to thicken the inside skin of the wall to make the foundations comply or ask an engineer for calculations.

PART FIVE
Mainly for Consultants/ Conclusion

25 Preparing to meet the client

What equipment should you take with you when carrying out your survey? I have included a list which covers most eventualities:

(a) Stout A4 clipboard.
(b) Adequate supply of paper. (**Note**: Some people prefer using graph paper as this ensures that sketches are made to scale.)
(c) Compass for checking north point.
(d) Variety of coloured pens. (This makes it easier to distinguish dimensions from the outline of the building.)
(e) Digital camera. (This saves a great deal of time for both the draughtsperson and for the surveyor when 'on site'. When preparing the plan, you are immediately reminded of what the property looks like and can pick up any items that you may have forgotten to note on your sketches such as rainwater pipe or, say, a balanced flue position.)
(f) Two metre folding surveyor's 'staff' (I find that a double-sided metric is best.) and/or laser measure.
(g) Manhole keys (heavy and light duty recommended); screwdriver and crowbar (for lifting manhole covers).

26 Meeting the client

GENERALLY

If you are just reading this book for general interest, then this section will probably only receive your passing glance. If you do intend to use the knowledge and take on clients, it is essential that you realize the true implications of your actions. Once you take on your first client, a contract comes into being even on the smallest projects. Your client will treat you as a professional and will expect you to have the necessary expertise to do the job. It is a requirement of most of the well-known professional institutions that surveyors carry professional indemnity insurance. This does not come cheap. Not having insurance is both against the 'rules' and rather foolhardy because we are all human and mistakes can happen. If for some reason you were found negligent, then the costs involved could run into many thousands of pounds.

THE INITIAL ENQUIRY

Although advertising is effective, the majority of enquiries are generated by people who were satisfied with the service offered previously and other personal contacts. This basic principle applies to most businesses whether they be large or small. Obviously, the more satisfied customers that you have, the more likely your business is to thrive. To take matters one stage further, if you look after your clients, like as not, they will come back or recommend your services to others. However, there is always the brand new customer. In order to turn that enquiry into work you need to be able to prove to your prospective client that you provide both an efficient and competitive service. What is more, as most enquiries are by telephone, you have to be able to convince your prospective customer, in a matter of a few minutes, that you can provide the service that he or she requires. In order that you can respond sensibly, you must try to anticipate what information that a client requires.

So what is your prospective client going to ask? The three most common questions that they are likely to ask are

(a) Can you carry out the service needed?
(b) How much do you charge?
(c) When can you meet to discuss their proposals?

HOW MUCH DO YOU CHARGE?

On the assumption that the answer to question (a) is 'Yes', and that you can meet your client you still have to answer question (b). In some cases, answering that question can be left until you meet. However, most clients want a 'budget'. So how much is a professional service of any sort worth?

It is a question of market demand and competition. It is easy enough to find out what others are charging. My only advice is do not undercut them too severely. It is easy to obtain work by undervaluing your own worth. There are factors which should affect your charges though. Before you can give your prospective client a 'budget' cost you need to know the following:

(a) Where do they live? They could live five minutes away or over two hours drive, and time is money.
(b) What do they want? Your client might just want a single-storey extension, two-storey or a garage.

TALKING THE JOB THROUGH

Once you have arrived at your client's house or the site of the proposed works, your client will obviously describe what he or she wants. Make sure that you note down all the important points. Do not rely on your memory. The important points to remember are

(a) Agree scope of work. A useful form to carry with you that covers most of the important items to discuss/note whilst on site is shown in Fig. 26.1.
(b) Tell the client what you do and what you don't do.
(c) Ask the client to sign a confirmation of instructions form at the time of survey. Obviously, with established clients this might cause offence but with a new customer it should not, as long as you are tactful. He or she is ordering work and you need a confirmation. If you obtain a signed order then a dispute over your agreed fee is unlikely. I have reproduced below a standard form that I carry with me when I visit a client. This is printed out on one of my letterheads (Fig. 26.2).
(d) Confirm your instructions. When sending the copy plan through to the client for approval, I normally send it with a letter on the lines of Fig. 26.3. This letter is then accompanied by a confirmation of instructions (Fig. 26.4).

Another useful form is one called 'What Do I Do Now?' (Fig. 26.5), which I send to clients to keep them updated.

STANDARD SURVEY SHEET

1. Name of Client ....

2. Address of Client ....

3. Owner Y/N

4. If tenant:

    address of owner..

5. Does the alteration affect neighbour by having footings projecting into the next door garden? Y/N

    Is a letter needed? Y/N

NB If you build the extension on the boundary the footings will be under the neighbour's garden. This is acceptable but only if the neighbour agrees (preferably in writing).

6. Name and address of neighbour ....

7. Instructions

    A. EXAMPLE CONSTRUCT REAR EXTENSION TO KITCHEN.

    B. EXAMPLE BLOCK UP OLD SIDE DOOR.

    C. ....

    D. ....

    E. ....

    F. ....

    G. ....

8. Where are drains/manholes? ....

9. Has the garden got over 50% of area left after building extension? Y/N

10. Has the house been extended before? Y/N

If yes explain to client that if existing extension was built after 1948, the cubic capacity counts towards 70/50 m^3.

11. Are there any trees close to the proposed extension? Y/N

Fig. 26.1 Standard survey sheet.

'ON SITE' FORM FOR CLIENT'S SIGNATURE

I confirm that the Conditions of Engagement have been explained to me and that I have been handed a copy for my retention. I accept the estimate figure of £ . . . . . for the service required and understand that this figure will not necessarily be the final cost in the matter.

I enclose my deposit of £ . . . . . and in so doing formally request that you commence the services as outlined.

I understand that I will receive a formal confirmation of instructions and a copy of the plans prior to submission to the Local Authority. Should there be any minor amendments that I require to the plan I understand that these will be made at my request at no extra charge as long as these amendments are made prior to submission.

I understand that before the formal submission can be made I must forward the Local Authority fees as requested in the Confirmation of Instructions.

SIGNED ....

NAME..

ADDRESS ....

POST CODE.................................

DATE ....

TELEPHONE ....

Fig. 26.2 'On-site' form for client's signature.

Dear Mr and Mrs Jones,

PROPOSED WORK AT YOUR HOUSE

I have pleasure in enclosing a copy of your plan as promised, a copy of my confirmation of instruction and a further copy of my Conditions of Engagement.

Should the plan require any amendments I would appreciate it if you would contact me as soon as possible so that valuable submission time is not lost.

It is essential that you check your plan over carefully, in particular the major dimensions, and ensure that everything is as instructed. I have prepared your plans based upon dimensions taken on site to 'Natural Boundaries' (e.g. fence lines).

If you are in any doubt regarding dimensions indicated on our plan then I would be obliged if you could contact me as soon as possible.

As agreed, my fee does not include for on-site supervision of the works.

Would you please forward the Local Authority fees as soon as possible, and I will then be able to submit the application on your behalf.

I will be forwarding my fee account to you for your attention under separate cover.

Yours sincerely,

A. R. Williams

Fig. 26.3 Covering letter for client's approval of plans.

Confirmation of Instructions

Client: Mr B. Jones

Address: Somewhere, London

Service: Prepare plans to client's approval and submit to Local Authority. No on-site supervision allowed. Agreed fee £ . . . . . (excludes planning and building regulation fee charges; any engineering fee charges which are or may be required by the Building Control Officers; on-site supervision of the works; and VAT of £ . . . . .).

Please forward a cheque for £ . . . . . for Building Control and a cheque for £ . . . . . for Planning. (Please make both cheques payable to Somewhere Borough Council and not to us.)

Note: Our fee account will be forwarded to you under separate cover. The Building Control and Planning charges are to be sent to this office prior to submission of the application. Should the Building Control fee be considered too low or too high, we will notify you and the details of credit/extra charge will be supplied.

Fig. 26.4 Confirmation of instructions.

WHAT DO I DO NOW?

1) EXAMINE YOUR PLANS

Please make sure that you are happy with the plans. Whilst we do try to attain the highest standard, we are only human. If there are amendments needed, please let us know immediately and we will put matters right, if we can.

Should the plan require any amendments we would appreciate it if you would contact us as soon as possible so that valuable submission time is not lost.

It is essential that you check your plan over carefully, in particular the major dimensions, and ensure that everything is as instructed.

We have prepared your plans based upon dimensions taken on site to 'natural boundaries' (e.g. fence lines). If you are in any doubt regarding dimensions indicated on our plan then we would be obliged if you contact us as soon as possible.

2) SEND US THE LOCAL AUTHORITY FEES

On the Confirmation of Instructions we have listed the Local Authority fees that we need to send your application to the Council. Please let us have these fees by return.

3) CAN I START WORK?

NO – we will submit your plans to the Council – it will take several weeks before this has been sorted out. We will send you your Planning/Permitted Development *AND* Building Control details once we receive them.

DO NOT START WORK UNTIL YOU HAVE HEARD FROM US

4) WHAT ARE WE DOING?

By law you need to obtain permission of the Local Authority before you build an extension. Normally this involves two departments which are:
1) Planning
2) Building Control
Once we have received your cheque/cheques for the Local Authority we will send your plans to the Council.

It is unadvisable to start work until these approvals have been received.

5) NOTES

(a) THE DRAWING/SUBMISSION FEES DO NOT INVOLVE ON-SITE SUPERVISION.

(b) COPIES OF CONFIRMATION OF INSTRUCTION AND CONDITIONS OF ENGAGEMENT ENCLOSED.

Fig. 26.5 Form – 'What Do I Do Now?'

27 Conclusion

GENERALLY

At the rear of the book, I have included:

(a) checklist of items not addressed or often missed off plans
(b) a standard specification
(c) building construction terminology.

These will hopefully be of help if you decide to prepare your own plans. I have tried to cover a large subject in a very small book. Hopefully, I will have succeeded in providing all the necessary basic background information.

However, regulations change and sometimes interpretations of regulations change so you must ensure that you keep up to date by studying the technical press and noting new problems as they occur. As no book can cover every eventuality, I have listed below some useful sources of information (some that I have already mentioned elsewhere in the book). Obviously, some of these links may change in time and you will have to locate the up to date Website.

Useful Weblinks

Approved documents
http://www.planningportal.gov.uk/england/professionals/en/1115314110382.html
Timber/Trada
http://www.trada.co.uk/
www.sipbuildltd.com
NHBC Standards
http://www.nhbcbuilder.co.uk/Buildersupportservices/TechnicalStandards/
Bricks
http://www.brick.org.uk/
http://www.ibstock.uk.com/
Blocks
http://celcon.co.uk/
http://www.thermalite.com/

Beam and block flooring
http://www.bison.co.uk/Beam-and-block-floors.aspx
Roofing/External plumbing/drainage
www.sipbuildltd.com
http://www.marley.co.uk/
http://www.marleywaterproofing.com/
http://www.redland.co.uk/
http://www.sandtoft.co.uk/
http://www.pekingslate.com/
http://www.icopal.co.uk/
http://www.hunterplastics.co.uk/
Insulation
http://www.knauf.co.uk/
http://www.celotex.co.uk/
http://www.insulation.kingspan.com/
Kitchen planning
http://www.diydata.com/planning/kitchen_design/kitchen_design.php
http://www.almostimpartialguide.co.uk/kitchens/layouts.htm

A FINAL WORD FOR WOULD-BE CONSULTANTS

As indicated in the previous chapter, if you are reading this for general interest, or hope to prepare plans for your own/family use, then matter such as insurance will probably not apply. However, if you are intending eventually to prepare plans for anyone other than direct family/close friends, then you must actively consider the legal implications of your actions. Although professional indemnity insurance is expensive, not having adequate insurance is foolhardy.

Appendix A Checklist of items not addressed or often missed off plans

(a) Under the 1976 Regulations a minimum floor-to-ceiling height of 2.30 m was required. The new Approved Documents do not make any requirement for a specific floor-to-ceiling height, except over staircases and under beams (2.00 m minimum and slightly less in loft spaces), and I suppose that if you wanted to design an extension with a ridiculously low floor-to-ceiling height the authorities might be hard-pressed to stop you, but I always indicate a minimum headroom of 2.30 m where possible as this is a sensible minimum height for rooms in a dwelling.

(b) The ground floor level/oversite concrete level should not be lower than outside ground level unless special waterproofing precautions have been taken.

(c) The DPC in the walls should be minimum 150 mm above ground level.

(d) The floor levels in the extension should be shown to be at the same level as the existing. Steps in floors should only be incorporated if absolutely necessary.

(e) Opening vents in windows. Mark the elevations so the Building Inspector knows which casements open (Fig. 1.1 shows details).

(f) Windows should be described as double glazed.

(g) Ventilation to habitable rooms. (Appendix B, clause 6 gives details.)

(h) When building across a driveway with a garage and there is no other access around the house, try to provide a rear door at the back of the garage (otherwise rubbish will have to be taken through the house).

(i) Windows in an existing house which open into a proposed attached garage should be blocked up to prevent a fire hazard.

(j) 'Rule of thumb' drains should fall to a gradient of 1:40 for 100 mm drains and 1:60 for 150 mm drains (however, note new formulas in Approved Document H).
(k) Where drains and sewers are close to proposed new foundations ensure that foundations are low enough not to impose any load on the drain or sewer. If a deep drain is near to a wall the trench must be backfilled with concrete.
(l) Generally, returns to windows should be at least one-sixth of the window opening it abuts (where between windows, one-sixth of combined window). No opening should exceed 3.00 m. If it does, engineer's calculations will be required.
(m) Between attached garages and houses provide half-hour fire doors complete with suitable frame having 25 mm rebates, door closer (e.g. two Perko door closers or one Briton closer). Provide 100 mm step to prevent petrol coming into dwelling in the case of fire.
(n) When specifying windows, doors, baths, kitchen fittings and so on, obtain catalogues (keep them in a filing system for frequent use). Particularly useful catalogues are Magnet and Southerns, and Boulton and Paul. Avoid specifying non-standard sizes.
(o) Cavity walls: your plans should be drawn in such a way that it is obvious that cavity walls are being used, by drawing in the cavity not only on the sections but also on the small scale plans.
(p) Cavity closing: see Appendix B clause 21.
(q) Garage walls/porch walls: a half brick (114 mm thick) wall can be used in garages and porches, but piers as specified in Approved Document A must be provided.
(r) Roof insulation and ventilation: pitched (tiled and slated roofs) of the cold type must be ventilated to the outside air so that moisture is 'sucked off'. Note the airway sizes shown in earlier chapters.
(s) Soakers and flashings at abutments: these are specially made pieces of lead (or approved substitute) which fill the gap between the roof tiles and an abutting wall and must be shown on the plans.
(t) Check drainage and locate manhole. It is a common fault to forget to provide drainage details on the plan.

Appendix B A standard specification

The following is a typical example of a standard specification.

COPYRIGHT

(1) This plan is copyright and is not to be produced or used for construction purposes without permission. It is only for use as a Planning and Building Control document.

THE PARTY WALL ACT

(2) From the 1 July 1997 any person intending to carry out works affecting party walls, or carrying out excavations for foundations adjacent to a party wall or in a position where such excavations are likely to affect a neighbouring property will be required to serve a notice on adjoining owners prior to commencement of work.
(3) Prior to commencement of works, the contractor must ensure that such notices have been given and that all formal procedures required under the act have been fulfilled.

BUILDING CONTROL REQUIREMENTS

Generally

(4) These plans shall not be acted upon until they have been approved in accordance with clauses 13 and 11(l)(b) of the Building Regulations. Should the owner or builder commence work without the above approval they do so at their own risk. All building construction is to comply the current Building Regulations and all Approved Documents. **All elements of structure shall have minimum 1/2 hour fire resistance**.
(5) In addition, all materials and building construction is to comply in all respects with the requirements of the National House Building Council's publication *NHBC Standards* (current edition). All goods and materials unless otherwise specified shall be in accordance with

the latest British Standard and Code of Practice current at the date of tendering.

(6) **Ventilation to rooms**

Room	Purge (rapid) ventilation (e.g. opening windows or doors) percentage assumes window/door opens more than 30 degrees	Background ventilation	Intermittent extract ventilation
Habitable room	5% of floor area	5000 mm^2	
Kitchen	5% of floor area	2500 mm^2	30 L/S to hob or 60 LS if elsewhere
Utility room	5% of floor area	2500 mm^2	30 L/S
Bathroom	5% of floor area	2500 mm^2	15 L/S
Sanitary accommodation (Separate from bathroom)	5% of floor area Mechanical extract of 6 L/S	2500 mm^2	

(7) **Part L1B of the Building Regulations**: Unless the main design drawings adopt another optional approach, the maximum *U*-values are

Element	*U*-value (W/m^2 K)	Means of achieving
Pitched roof – Insulation at ceiling level	0.16	Roof insulated with two layers of crown Loft roll 40, 100 mm between ceiling joists and 200 mm laid over the joists.
Pitched roof – Insulation at rafter level	0.20	80 mm Celotex GA308DZ between the rafters and 50 mm GA3050Z fixed to underside of rafters (400 centres), with 12.5 mm plasterboard and skim ceiling below. Or 130 mm Crown Rafter Roll 32 between the rafters and 36/9.5 Polyfoam Linerboard as internal lining. A 50 mm ventilated void should exist above the Rafter Roll (rafters have to be at least 180 mm deep).
Flat roof with intergral insulation	0.20	Warm deck: 105 mm Kingspan Kooltherm K11 Roof board bonded to vapour control membrane and finished with bitumen built up roofing sytems. Or Ventilated roof: 100 mm Celotex Exra-R XR3000 between joists (at 400 mm centres) an 40 mm Celotex Tuf-R GA3000 to underside of joists.

Dormer walls: Timber framed with tile hanging or PVC cladding externally	0.30	75 mm Polyfoam Rafter squeeze between studs and 36/9.5 mm polyfoam linerboard internal lining. Or 100 mm Mineral wool batt between studs with 3/9.5 mm polyfoam linerboard internal lining. Or 65 mm Kingspan Thermawall TW51 between studs and a 32 mm Kingspan Kooltherm K18 insulated dry lining internally.
External walls: Masonry walls of cavity construction with brick outer leaf	0.30	Total fill with 85 mm Crown Drytherm and inner leaf of 100 mm standard aircrete block ($\lambda = 0.16$ Celcon, thermalite or Durox) with and internal finish of plasterboard on dabs. Or Partial fill with 45 mm Kingspan TW50 (a minimum 25 mm residual cavity must remain adjacent to outer leaf) inner leaf of 100 mm of standard aircrete blocks ($\lambda = 0.16$) with and internal finish of plasterboard on dabs. Or Partial fill with 50 mm Cellotex Tuff-R 3000 (a minimum 25 mm residual cavity must remain adjacent to outer leaf), inner leaf of 100 mm of standard aircrete blocks ($\lambda = 0.16$) with and internal finish of wet plaster.
Wall between garage and the house	0.30	215 mm Lightweight aircrete (Celon Solar/Thermalite turbo) lined with 37.5/12.5 Kingspan Kooltherm K18 Dry-lining board.
Ground floor Note the thickness of the insulation required will vary dependent on the shape and size of the floor (the P/A ratio)	0.22	Traditional solid concrete floor construction insualted with 75 mm + 35 mm Polyfoam Floorboard standard insulations Or 140 mm Jabfloor 70 Or 80 mm Celotex Tuff-R
Floor over garage	0.22	200 mm Rocksill flexible slab between joists Or 100 mm mineral wood quilt between joists with 80 mm Celotex Tuff-R 3000 above that.
Windows roof windows and roof lights (**Note**: Windows can be rated using BFRC Rating system A to G – A is best)	1.80 ***	Or window energy rating band D Or centre pane value of 1.2

contd.

Element	*U*-value (W/m^2 K)	Means of achieving
Doors with more than 50% of internal face glazed.	2.2 ***	Or centre pane value of 1.2
Other doors	3.00 ***	
		*** Unless the thermal insulation of the windows/doors is improved by the use of such products as Optitherm the areas of windows and roof windows and doors should not be greater than 25% of the floor area of the extension plus areas of any windows and doors no longer exposed as a result of the extension works.

Foundations

(8) Dimensions given on foundations are only indicative for normal soil conditions. Should it be necessary to provide raft foundation or other construction the builder shall contact the surveyor as soon as possible.

(9) All foundations to be taken down to good bearing strata and in accordance with local authority requirements. Concrete strip foundation shall comply with the the requirements of NHBC Standards 2006 and be minimum GEN1. If the concrete is to be laid in aggressive ground the contractor will comply with Part 2.1 Tables 4a and 4b of NHBC Standards. Unless detailed otherwise on the main drawings, foundations shall be taken down a minimum depth of 1.00 m from ground level and the concrete foundations shall be 600 mm wide × 200 mm thick minimum size. Should the Building Inspector instruct that differing sizes or mixes of concrete to be used, the contractor is to comply with such instructions.

(10) Brickwork/blockwork in foundations shall comply with Part 5.1 of NHBC Standards.

Damp-proof courses

(11) Provide adequate DPC's to new brickwork. The DPC's to be a minimum of 150 mm above adjoining ground level. Returns to windows and door openings in new cavity brickwork to have vertical DPC.

Solid floor slabs and finishes

(12) The floor slabs, oversite concrete or rafts if applicable, shall not be laid lower than the adjacent ground level. If existing levels are inconsistent with this the contractor must lower adjacent levels accordingly.

(13) Ground floors to simple domestic properties the floor construction shall comply with Table 1 and shall be constructed as follows: approved 50 mm cement and sand screed (or approved flooring asphalt), laid on 150 mm concrete floor slab GEN 1on, 1200 gauge Visqueen laid over hardcore covered with a sand blinding. Unless thermal blockwork (Thermalite/Celcon or Durox) abuts the edge of the floor, approved insulation should be turned up at edges to prevent cold bridges. The Visqueen should also be turned up at edges of the floor slab to link up with the DPC in the walls.

Block strutting

(14) Block strutting to be provided, where joist ends bear onto steelwork, where joist ends ear onto intermediate walls, where joist ends are supported on joist hangers and where joists span over 2.5 m in accordance with NHBC regulations.

External walls

(15) The external walls to new habitable areas shall have a minimum *U*-value shown in Table 7. All bricks and blocks used in the construction shall comply in all respects with NHBC Standards. Where exposed to view all returns in facing brickwork shall be in facings.

(16) Cavities shall contain stainless steel wall ties 5 per m^2 spaced at 750 mm centres horizontally (max) and 450 mm vertically and additional ties at openings.

(17) Where cutting and toothing is not possible 'fir fix' stainless steel connections shall be used.

(18) The contractor must ensure that ends of beams are adequately seated on suitable padstones and that lightweight blockwork is not over stressed.

(19) The heads of all cavity walls and returns around door and window openings shall be closed with thermal blockwork (incorporating proprietary insulated cavity closer to prevent cold bridging) if the inner blockwork does not have the required thermal rating recommendations and incorporated a DPC.

Windows

(20) All new windows are to be installed by a FENSA approved contractor.

Straps

(21) Roof joists, rafters and floor joists shall be strapped in accordance with the Building Regulations using 30 mm × 5 mm galvanized mild steel straps and shall comply with NHBC Standards and be fixed at 1.2 m centres to provide vertical and horizontal restraint.

Flat roofs

(22) Flat roofs shall be of the warm deck type and have a minimum *U*-value specified in Table 7 above. The builder shall comply with NHBC Standards Chapter 7.1. All roof finishes shall carry a 15-year guarantee.

Pitched roofs

(23) Mortar used on the ridge and verge shall match the colour of the tile.

(24) The sarking felt shall be reinforced underslating 'Andersons Twinplex' type IF or other equal and approved, fixed to the SW rafters with No. 2 galvanized nails size 20 mm into each rafter. The felt is to have 150 mm headlaps, and is to lap 50 mm into the gutter.

(25) Where an existing roof is re-covered, the contractor must comply with Approved Document C.

(26) All roof pitches shown on the drawings are assessed pitches based upon photographic records taken from ground level and are therefore only indicative of likely pitch (angle of slope). The contractor must calculate the correct pitch from dimensions taken on site.

(27) Where a new roof section abuts adjacent tiling the new tiles should match the existing tiles in type colour and pitch. (Should the existing tiling not be at recommended pitches the surveyor should be notified immediately.)

(28) Where Marley Wessex or Marley Modem tiles are specified, the tiles will be smooth and must not be laid lower than 15 degrees and 17.5 degrees respectively.

(29) Where it proves impossible to lay the tiling at manufacturers pitches on any roofing system, the Marley specification 'Laying Roof Tiles below recommended pitches' shall be followed. In brief the construction below the tiles shall be as follows, 38 mm × 25 mm treated battens on single layer BS747 iF felt, on 36 mm × 19 mm treated counter battens, on one layer BS 747 3B felt continuously bonded with 115/15 bitumen on one layer glass fibre based felt, random tacked on and including 12.5 mm external grade plywood.

Leadwork

(30) The contractor shall install all necessary soakers, flashings, aprons, and the like at all abutments sufficient to prevent water entering the building.

Windows/doors and finishings

(31) The contractor shall allow in his price and install all plasterwork, screeds, skirtings, doors, windows, architraves, window boards and so on.

Sound insulation

(32) Upper floor of two-storey extension should have a minimum the 22 mm tongued and grooved chipboard. The tongues and grooves should be glued with an approved glue. The chipboard shall be Type P5 moisture-resistant chipboard. Inside the floor, there should be 100 mm mineral wool with a density of 10.5 kg/m^2. The ceilings are to be lined with Gyproc Wallboard 10 (or similar).

Electrical work (Part P)

(33) All electrical work covered by Part P (Electrical Safety) must be designed, installed and tested by a person competent to do so. This person must be registered with an authorized self-certification scheme (e.g. BRE Certification, BSI, Elecsa, NICEIC, or NAPIT Certification). Prior to completion the Council should be satisfied that Part P has been complied with. This may require an appropriate BS 7671 electrical installation certificate to be issued for the work by the competent person.

(34) Light switches and electrical socket outlets should be installed between 450 mm and 1200 mm above floor level for easy reach. Typically socket outlets should be at 450 mm above floor level and light switches 1.00 m to 1.20 m.

(35) All new lighting outlets shall be fitted with lamps with a luminous efficiency greater than 40 lumens per circuit watt.

Smoke/fire alarms

(36) There should be a self-contained smoke alarm within 7 m of the doors to rooms where fire is likely to start and within 3 m of bedroom doors. Self-contained smoke alarm to be fitted centrally to landing at least

300 mm clear of existing light fittings and wired to the mains on a separate fused circuit in accordance with Institution of Electrical Engineers (IEE) regulations. Alarms to be certified to BS 5446.

(37) Multiple smoke alarms should be interconnected so that detection of smoke on one sets off all alarms.

(38) Provide a heat detector to the kitchen and connect to the other smoke alarms.

Plumbing/central heating/heating controls

(39) All new and replacement natural gas and LPG boilers are required to have a minimum SEDBUK (Seasonal Efficiency of Domestic Boilers in the UK) rating of 86 per cent.

(40) Oil fired boilers must have a minimum SEDBUK rating of 85 per cent.

(41) Exceptional circumstances permitting the installation of a non-condensing boiler, the installer must complete an 'Assessment form' using the procedure described in the document 'Guide to the condensing Boiler Installation Procedure for dwellings' (ODPM 2005). The declaration should be retained by the householder as it may be needed when the property is offered for sale.

(42) New hot water radiators should be fitted with thermostatic radiator valves or be controlled by room thermostats.

Drains

(43) Drains shall be Hepseal or Hepsleeve flexible jointed 100 mm clay pipes with 150 beds and surround laid in accordance with Hepworth recommendations to falls of 1:40. Encase all drains under extension with 150 mm concrete.

(44) Lintels to be provided in substructure walls where drains pass through foundations as per NHBC Standards.

(45) New inspection points, chambers and manholes shall be constructed in accordance with the schedules in the Building Regulations. New manholes (where provided) shall comprise 150 mm concrete bed with benching to channels, 255 mm class B engineering bricks in cement mortar 1:3, 150 mm reinforced concrete cover slab and mild steel cover – All to the total satisfaction of the District Surveyor.

The builder/building contractor

(46) The terms 'builder' or 'contractor' shall mean the person responsible for the construction of the works.

(47) The contractor shall ensure that a responsible person is on site during normal working hours to take instructions.

(48) The contractor is advised to visit the site, prior to quoting and to make due allowance when preparing his estimate for access, availability of labour, plant and all things necessary for the construction of the works. No claim will be accepted for want of knowledge at a later date. Give all notices to local authority and public undertakings and pay all fees and charges. (**Note**: The building control fees for on site inspection of the works will be paid by the client unless agreed otherwise.) The contractor shall include for all costs arising from compliance with all Statutory Orders, Regulations, Building Regulations, Bye-Laws and any Acts of Parliament.

(49) The contractor shall protect the premises during the execution of the work against all damage or vandalism and shall provide tarpaulins and all other necessary coverings, and take adequate precautions to keep new and existing work free from damage by inclement weather during the progress and clear away on completion. The premises are to be secure at the end of each day's work, and the contractor must reinstate at his own expense any damage caused by neglect in protecting the building.

(50) The contractor must furnish the local authority with notices of commencement of work and stage of completion and must liaise with water, gas, electricity and British Telecom as necessary and comply with their requirements. The contractor will be required to maintain and protect all gas and water pipes, electricity cables, sewers etc, and other public property or property of the local authority or public utility company which may be encountered during the progress of the works and he shall be responsible for and properly make good any damage to the same to the satisfaction of the authorities concerned. No part of the work shall be sublet to other persons unless the written authority of the owner and surveyor is obtained. All materials, appliances, fittings, and the line must be obtained from sources approved by the owner and surveyor with reasonable samples of materials to be used in the work, samples if approved shall become a standard of quality.

(51) The contractor is to carefully check the boundaries of the site and is not to build on land not owned by the client without obtaining the neighbours consent.

(52) The contractor must indemnify and insure the owner for any damage to persons and/or property for the sum of not less than £500,000 and the contractor will be held liable for any damage or nuisance to, or trespass on the adjoining property arising from or by reason of the execution of the work, and he must take all necessary steps to prevent any such trespass or nuisance being committed.

(53) No trial holes have been taken on site and the builder must acquaint himself with the ground conditions of both the site and adjoining areas.
(54) Comply with NHBC Practice note no. 3 – root damage by trees, siting of dwellings and special precautions.
(55) Allow for keeping the works clear of rubbish during the currency of the contract and remove from time to time all debris as it accumulates and leave clear and tidy on completion to the satisfaction of the owner.

Drawings/dimensions

(56) The drawing and specification is to be read as a whole; if any details whatsoever are not clearly shown or specified the contractor is to ask for instructions, and if any work be wrongly done, it shall, if the surveyor so directs, be removed and done again at the contractor's expense.
(57) Site copies of the drawing must be available on the site during the progress of the works for inspection by the LBO.
(58) All dimensions given whether figured or scaled are to be physically checked on site by the contractor prior to commencement of work and the contractor will take responsibility for same.

Appendix C Building construction terminology

Aggregate Broken brick, gravel and sand which forms a major part of materials such as concrete and mortar.

Air brick A perforated brick built into a wall to provide ventilation either to a hidden void (e.g. under a timber floor) or a room.

Asbestos Fibrous mineral. Airborne fibres are a known health hazard.

Ashlar Tightly fitting, square cut building stones.

Asphalt Thick bituminous coating applied hot to waterproof flat roofs, basements etc.

Back inlet gulley (BIG) Ground level inlets which are connected to the drainage system into which waste and storm water discharges. Unlike a standard gulley, a BIG has an additional inlet (or inlets) to ensure that connections to above-ground drainage can be easily made.

Balanced flue A duct through a wall which takes air to a boiler from outside and expels waste gases. Usually wall-mounted externally behind modern boiler.

Balustrades Protective handrails and spindles for staircases and balconies.

Barge board Similar to a fascia board but runs up the gable of a house covering up exposed roof timbers.

Battens Thin timber strips to which tiles and slates are fixed.

Benching Sloping concrete at base of drainage manhole.

Binder Roof timber running over ceiling joists to provide stiffness.

Birdsmouth Cut in roof timber to join strut at angle to purlin, wall-plate or other structural timber.

Blockwork (building blocks) Precast concrete blocks approximately 18 in × 9 in (450 mm × 225 mm). Generally cheaper to build than brickwork. Now a high quality factory-produced product, usually made of thermally insulating material and manufactured by a large number of companies (e.g. Celcon, Thermalite). The term breeze block should not be used as these blocks are no longer made.

Bonding Method of laying bricks to a regular pattern, for example English bond, Flemish bond, English Garden wall bond, stretcher bond.

Borrowed light Window in interior wall transferring light from outer window.

Brick Bricks come in various types (e.g. facings (expensive but look nice), commons (cheaper than facings, usually used in foundations), engineering (very hard dense bricks which are used occasionally where high loading or water-resisting qualities needed)). Distinct from a building block, usually made of burnt clay but can be made from concrete or calcium silicate. In very approximate terms the modern coordinating brick used in the UK is usually 9 in × 4.5 in × 3 in high (225 mm × 112 mm × 75 mm) and there are approximately 60 bricks per square metre of half-brick wall. For more details see Chapter 8.

BS British Standard.

Building Regulations Statutory Local Authority control over building works.

Casement Part of a window.

Cast *in situ* Material cast on site in formwork – usually reinforced concrete.

Cavity insulation A cavity in a cavity wall can be insulated using expanded polystyrene or glassfibre bats or blown insulation. Used to increase the thermal efficiency of a wall. Only purpose made, water-repelling insulation should be used (e.g. Dritherm).

Cavity tray A type of damp-proof course which steps across a cavity wall to ensure that water in the cavity drains towards weep holes in the outer skin of the cavity wall.

Cavity wall A warm, dry wall used in UK since the 1920s as an alternative to the solid wall. Usually comprises an outer skin of facing brickwork with an inner skin of thermal blocks. Between the two skins there is usually 50 mm gap which stops water passing from the external face to the internal face of a wall. This cavity can be nowadays filled with approved cavity insulation.

Cavity-wall tie (or wall tie) Purpose-made tie which has been galvanized or made from stainless steel to link the two skins of a cavity wall together. There are many patent types on the market but the standard types are 'butterfly' or 'vertical twist'.

Ceiling joist A joist that only carries the weight of the ceiling and not a floor.

Cement The term usually refers to ordinary Portland cement as used in the making of concrete.

Cesspit Non-mains drainage using sealed tank emptied periodically by council.

Cheek Side face of dormer etc.

Codes of Practice Various non-statutory recommendations for use of materials.

Collar Roof timber running between opposing rafters to prevent spread.

Condensation Water deposit on any surface when critical dew point is reached.

Consumer unit Modern electric switch box containing fuses or circuit breakers.

Conventional flue Boiler chimney with boiler air taken from room.

Coping Brick, stone, or tile finish to top of parapet wall.

Corbel Projecting support on face of a wall.

Cornice Decorative plaster moulding at junction of wall and ceiling.

Cowl Shaped chimney pot used to prevent down draught.

Cut, tooth and bond A method of joining old brickwork to new by removing some of the old bricks and lacing in the new so that the new and old structure are adequately tied together.

Dado Lower part of internal wall approximately 1 m high below timber rail (dado rail).

Damp-proof course (DPC) Usually a strip of patent water-resisting material incorporated into a wall or around a window or door opening to prevent water penetration.

Damp-proof membrane (DPM) In modern construction usually a building film such as Visqueen sheet laid below floor slabs to prevent water penetration to the upper surface. The term is also applied to tar-based applications which serve the same function.

Door closer Usually used on fire doors to ensure that the door returns to its fully closed position in case there is a fire.

Dormer Window projecting out of roof slope.

Dragon tie A diagonal piece of timber or metal strap which is fitted internally across the corner of a hipped roof to prevent the roof spreading.

Dry rot A fungus (*Serpula lacrymans*) that destroys timber if the conditions are favourable. This rot, once established, will travel extensively and force its way through brickwork and plaster to infect new timber.

Eaves Lowest part of a sloping roof or the area under it.

Efflorescence Salt deposits on walls or roof tiles where dampness evaporates.

Fabric reinforcement Wire mesh reinforcement usually used in concrete slabs. Usually specified using a BS code (e.g. A142).

Façade Front elevation of building.

Fascia board A piece of boarding that supports the gutters or covers up exposed roof timbers.

Fillet Triangular sealing of joint between surfaces, generally cement mortar or timber (as in tilt fillet).

Flashing Lead, zinc, copper or patent strip covering junction of roof slope with wall or chimney.
Flaunching Fillet of mortar surrounding base of chimney pots.
Flight Straight run of stairs.
Gable Triangular upper part of end wall from eaves level to ridge (sometimes also referred to as 'pike' by bricklayers, but this is a regional term).
Gang-nailed trusses Modern prefabricated roof timbers fixed with plates (e.g. fink truss or monopitch) (Fig. 19.8 shows details).
Going Distance between risers in a staircase.
Gulley Ground level inlets which are connected to the drainage system into which waste and storm water discharges.
Gypsum plasterboard See plasterboard.
Hardcore Broken brick, broken stone, concrete etc.
Header Brick laid with short end exposed.
Herringbone strutting Cross-shaped layout of timbers nailed between joists as stiffening. It is now possible to obtain purpose-made metal struts (e.g. Expamet struts).
High alumina cement (HAC) Must not be used in structural work.
Hip Sloping edge joining two pitched roof slopes.
Honeycomb wall Bricks laid with gaps between to allow ventilation.
Hopperhead Funnel to collect water at top of downpipe.
Invert Bottom of manhole or drain.
Joist Timber support to roof ceiling or floor running parallel with ground.
Lintel (or lintol) Beam or patent steel beam or joist used over opening supporting construction above.
Made ground Potentially difficult sites infilled with hardcore or rubbish.
Mineral felt Common modern flat roof covering with fairly short life.
Monopitch Roof with only one slope.
Mullion Vertical member dividing panes in a window.
National House Building Council (NHBC) See Chapter 9 for brief details.
Nosing Outer top corner of step or sill.
Oriel window Upper floor window which is cantilevered into a projecting bay.
Outrigger Regional term (Merseyside in particular) usually used when describing a projecting rear extension to a terraced property.
Oversailing course Course of brick or stone projecting out from face of wall (continuous corbel).
Parapet Top of wall carried up above roof or balcony level.
Party wall Each owner owns half with rights in respect of the other half.
Piles Foundation of concrete columns sunk into ground.
Pitched roof Sloping (rather than flat) and covered with tiles, slates etc.
Plasterboard Gypsum plaster sandwiched between two sheets of stout paper.

Plinth Widening at base of wall.

Plumb Vertical.

Ponding Pools of water lying on a flat roof. If water is lying in patches deterioration can result.

Purlin Major roof beam supporting rafters which runs sideways across slope supporting the rafters.

Raft foundation Shallow flat reinforced concrete slab used as alternative to a strip footing.

Rafter Sloping roof timber supporting battens and tiles or slates.

Rendering A surface application of cement and sand (or similar mix) to external face of wall to provide waterproofing.

Retaining wall Supports ground behind and may provide support to structures.

Reveal Return face of corner at window and door openings.

Ridge Top edge of pitched roof.

Rise Vertical distance between stair treads.

Riser Vertical front of a step.

Rising damp Moisture passes up walls and through floors by capillary action.

Rolled steel joist (RSJ) Sometimes used as lintel over openings.

Sarking felt Underfelting used beneath battens and tiles on sloping roofs.

Sash Inner frame to window carrying glass, hence sliding sash windows.

Screed Smooth cement or asphalt finish to concrete floors.

Scrim Hessian-type material used to seal joints in plasterboard.

Septic tank Non-mains drainage using bacterial action to break down sewage.

Serpula lacrymans The Latin name for the dry rot fungus.

Sleeper walls Used beneath timber suspended ground floors to support sleeper plates and joists (e.g. honeycomb sleeper wall).

Soakers Lead or zinc angles between tiles or slates and flashings.

Soffit Underside of eaves behind fascia.

Soil stack (or soil and ventilation pipe – SVP). Main vertical drainpipe for WC and other waste water.

Soldier arch Bricks laid vertically on end at top of window or door opening.

Spalling Breaking of surface of tiles or bricks, often due to frost action in winter.

Sprocket Angled timber at foot of rafter to lift roof tiles or slates over gutters.

Stretcher Brick laid sideways, i.e. for single-skin or cavity work.

Strings Sloping boards supporting ends of treads to staircase.

Strip footing Strip of concrete buried in the ground to support a brick wall.

Stucco Architectural term to describe smooth cement rendering used as external finish to walls (also known as render or rendering).

Stud partition Timber-framed walls clad in plasterboard.

Subframe Outer part of window fixed directly to sides of opening.

Subsoil Material below topsoil on which foundations rest.

Sulphates Chemicals in the ground which can cause weakening of concretes especially below ground.

Sulphate-resisting cement Used as an alternative in concrete when subsoil conditions likely to destroy ordinary Portland cement.

Thermoplastic tiles Floor tiles made from thermoplastic resins.

Threshold Sill to an exterior door opening.

Timber-frame houses Built with load-bearing timber and (usually) brick face.

Torching Mortar applied to underside of slates or tiles (an old-fashioned, out-of-date construction technique).

Transom Horizontal window member separating panes (vertical is mullion).

Trap Bend in waste pipe prevents air from drain rising (usually found under sinks, wash-hand basins etc).

Trussed rafter See gang-nailed trusses.

Underpinning Excavating and inserting a new foundation under existing foundation.

Upstand Vertical face of flashing or soaker or concrete/timber projection.

Vapour check Barrier to prevent warm damp air entering wall or roof void.

Verge Edge of pitched roof at gable end.

Vertical damp-proof course Used at change in level and in basement and around to window and door openings or at a cavity closing.

Weep holes Allow water to drain from cavity walls and from behind retaining walls.

Wet rot Fungal attack to woodwork, especially exterior softwood joinery.
Not as serious as dry rot.

Appendix D 'Housing Extensions' Halton Borough Council

Halton Borough Council

House Extensions

Supplementary Planning Document
December 2006

Operational Director
Environmental and Regulatory Services
Environment Directorate
Halton Borough Council
Rutland House
Halton Lea
Runcorn
WA7 2GW

This guidance note should be read in conjunction with the relevant policies of the Development Plan.

Contents

1 Purpose

1.1 The purpose of the House Extensions SPD is to complement the Halton Unitary Development Plan (UDP), by providing additional guidance for anyone intending to extend or alter their house or erect a garage or other outbuilding to ensure that all developments:

a are of exemplary design quality and that any extensions do not spoil the character of the original dwelling, but relate closely to it and harmonise with the existing house in its scale, proportions, materials and appearance;
b protect residential amenity of neighbouring properties;
c protect and enhance the built and natural environment;
d preserve the essential character of the street and surrounding area;
e avoid the creation of dangerous highway conditions; and
f safeguard the provision of a reasonable private garden space.

1.2 By stating this purpose, the Council will seek to improve through its function as the Local Planning Authority any development proposal that does not provide for, or meet the principles encouraged and required within this SPD and the Halton UDP

1.3 These guidelines are intended to illustrate the criteria that will be applied by the council in assessing proposals for house extensions. They also provide advice for planning and designing domestic extensions in a way that will enhance the appearance of the dwelling whilst maintaining the character and amenity of the neighbourhood.

1.4 This SPD is also intended to encourage residents to follow the practical guidance it contains wherever opportunities arise whether or not formal consent is required.

1.5 It is important that each section is not read in isolation, as guidance provided in the whole of the document will be applied, as required.

APPLICATION OF THE POLICY

The definition of 'house' in the policy includes bungalows, but excludes apartments or maisonettes.

'Extension' means all additions to the house whether attached or not, and includes garages.

This House Extensions SPD also applies to:

- Houses that are listed buildings and buildings in Conservation Areas.
- Houses in the Green Belt.

However, due to the special characteristics of these areas, more stringent controls may need to be applied.

Exceptions may be considered for an extension to provide basic amenities or facilities at ground floor level for a disabled person. However, where possible, the extension should be designed to comply with the guidance. In circumstances where the guidance cannot be adhered to, acceptable proof of disability and a written statement justifying why an exception should be made shall be submitted with any application.

Please note that the diagrams used within this document are illustrative and are not drawn to scale.

2 Guiding Principles

2.1 This section sets out some of the documents that contain guiding principles that have been used to inform the general principles set out in this document.

By Design

2.2 By Design, Urban Design in the Planning System: Towards Better Practice, is a companion guide to National planning policy guidance. It does not provide policy, but encourages better design. By Design summarises the objectives of urban design as:

- **Character** – A place with its own identity
- **Continuity and enclosure** – A place where public and private spaces are clearly distinguished
- **Quality of the public realm** – A place with attractive and successful outdoor areas
- **Ease of movement** – A place that is easy to get to and move through
- **Legibility** – A place that has a clear image and is easy to understand
- **Adaptability** – A place that can change easily
- **Diversity** – A place with variety and choice

Securing the Future

2.3 Securing the Future: Delivering UK Sustainable Development Strategy, sets out the Government's sustainable development agenda. It sets the following guiding principles for sustainable development:

- Living within environmental limits
- Ensuring a strong, healthy and just society
- Achieving a sustainable economy
- Promoting good governance
- Using sound science responsibly

Building in Context

2.4 The belief underlying 'Building in context' is that the right approach is to be found in examining the context for any proposed development in great detail and relating the new building to its surroundings through an informed character appraisal. It suggests that a successful project will:

- relate well to the geography and history of the place and the lie of the land;
- sit happily in the pattern of existing development and routes through and around it;
- respect important views;
- respect the scale of neighbouring buildings;
- use materials and building methods which are as high in quality as those
- create new views and juxtapositions which add to the variety and texture of the setting.

3 General Principles for all Extensions

3.1 These principles apply to all extensions:

Design in relation to existing dwellings

3.2 An extension should relate closely to, and harmonise with the existing building in its scale, proportions, materials and appearance. In particular:

- The size of the extension should be subordinate to the size of the dwelling as first built.
- The external materials used shall closely match those of the existing dwelling in their design (see diagram 1).
- On prominent elevations, problems of bonding old with new brickwork on the same plane should be overcome by setting the extension back from the main wall of the dwelling.
- The roof of an extension should be pitched to match that of the existing dwelling. Flat roofs are not normally acceptable, except where they are a feature of the original dwelling house. (see diagram 2)

Diagram 1: Poor choice of external materials for the extension

- The windows of any extensions should be in line with existing windows and should match their proportions, size and design. (see diagram 2)

Diagram 2: Examples of side extensions – one with an appropriate pitched roof and well matched windows and one with an inappropriate flat roof and poorly matched windows.

How and where to extend

3.3 The choice of how and where to extend will depend upon a variety of factors. However, the Council's policies as set out in this booklet will be an indicator as to whether or not a particular proposal will be acceptable.

3.4 The position of the dwelling within its plot will be one of the most important considerations. Is there more space at the side or at the rear? How will the extension relate to the internal arrangement of the house?

3.5 For dwellings in substantial plots, there may be several options where an extension can be accommodated with little affect on neighbouring properties.

3.6 However, a basic principle to follow is that extensions should respect the style and character of the original house and not overwhelm it.

Effect on the street scene and the character of the area

3.7 Apart from its relationship to the existing house, an extension should not be visually detrimental to the existing character or appearance of the street scene or the surrounding area. In particular:

- Where a house is one of a group, similar in appearance and significant in the street scene, the effect of an extension to that house on the appearance of the group, as well as the individual house should be carefully considered.

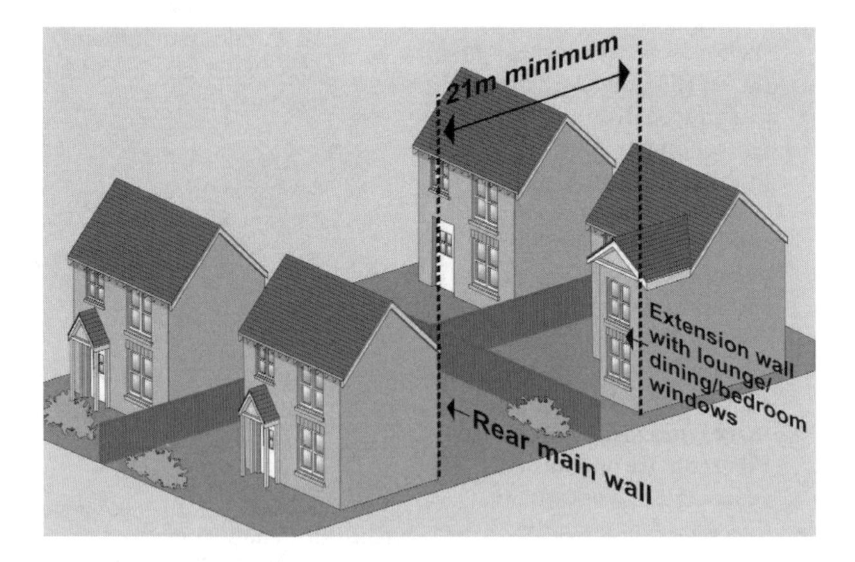

Diagram 3: 21 metres between facing principal windows

- An extension should respect any regularity and width of spaces between existing houses and the visual effect of these spaces when significant in the street scene.
- An extension should respect any regularity in the distance between the road and the frontage walls of existing houses when this distance is a significant factor in the street scene.
- An extension at the rear of a dwelling should not be so extensive in relation to the size of the rear garden or yard that the enlarged house would constitute overdevelopment of the site that would be out of character with the area.

Amenity of neighbours

3.8 An extension should respect the existing standard of daylight and privacy experienced by neighbours, in particular:

- Where principal windows will allow views to other principal windows of a neighbouring property, a minimum distance of 21 metres must be maintained. (see diagram 3)
- Where principal windows directly face a blank elevation, a minimum distance of 13 metres must be maintained. (see diagram 4)
  - Where the house concerned is more than

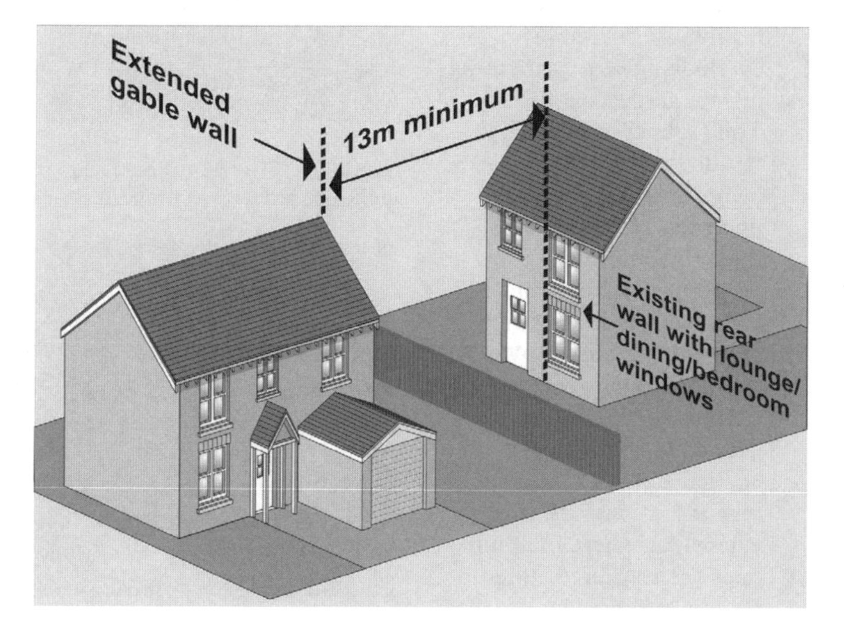

Diagram 4: 13 metres between principal windows and blank wall

two storeys, the Council will normally apply a further 3 metres for each additional floor in addition to the distances stated above. For example, a three-storey house will normally require 16 metres (13m + the additional 3m) between a principal window and a blank elevation or 24 metres (21m+3m) between facing principal windows.

- Any new patio area or balcony at first floor levels should not have the potential for an unacceptable degree of overlooking from any main window of a principal room in an adjacent house; nor for the direct sideways overlooking of neighbouring private garden or yard.
- The Council uses the '45-degree rule' to help assess impact upon the amenities of the neighbouring properties and to protect from overshadowing or obstruction, caused by large extensions on or close to the boundary. The code is principally applied to single storey rear extensions and side extensions where building lines are staggered. (see diagrams 5, 6 & 7)

Note: a principal window is a main window of a living room, dining room, conservatory or a bedroom.

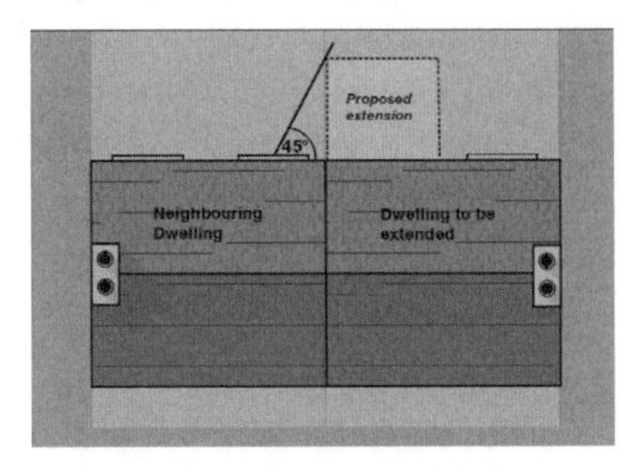

Diagram 5: The '45-degree Rule' applied to a semi detached or terraced property

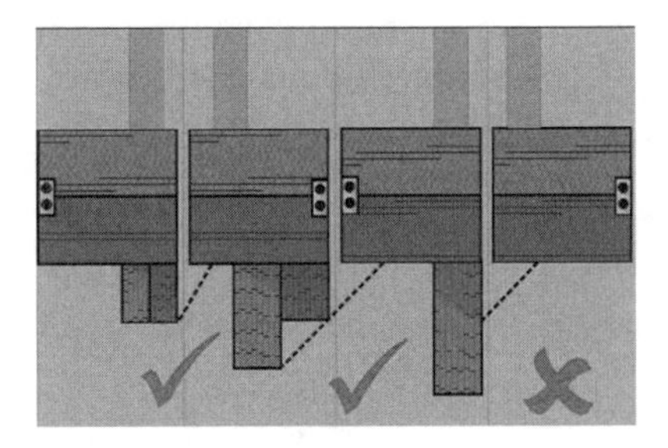

Diagram 6: The '45-degree Rule' applied to a detached property

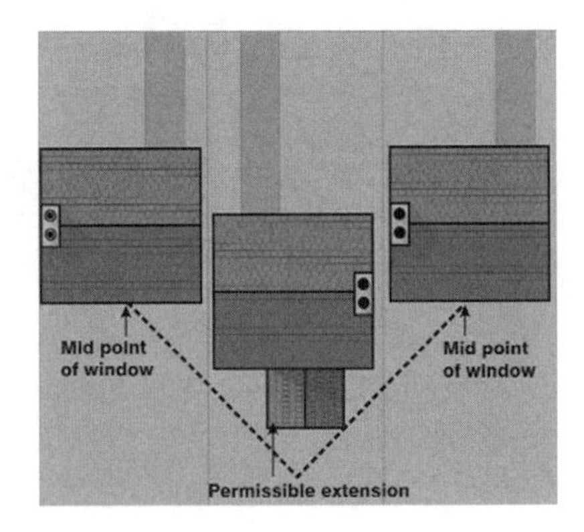

Diagram 7: The '45-degree Rule' applied to staggered properties

Building Control

3.9 Building regulation requirements should be taken during the design of any alteration or extension. More advice on the building regulations may be obtained from the Council's Building Control Division, contact details can be found in the Appendix.

4 Front Extensions & Porches

4.1 These principles apply to front extensions and porches:

- Front extensions should respect the existing property, and neighbouring properties, regarding design, size and siting.
- Any extension to the front elevation must be designed to harmonise with the existing property.
- Proposals should not result in the loss of existing parking provision.
- If the buildings on the street follow an established pattern or clear building line, front extensions are more likely to be considered to adversely affect the appearance of the street scene.
- Where planning permission is required, conversions of integrated or attached garages will not normally be allowed if two off road parking spaces cannot be provided.
- Porch extensions should also match the original design of the property.
- The height of the porch should not exceed the sill height of the first floor windows.

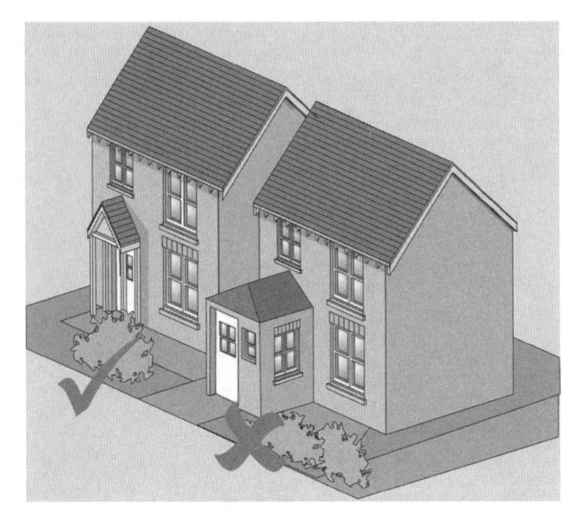

Diagram 8: Inappropriate porch extension.

5 Side Extensions

5.1 To avoid terracing and / or an unbalanced effect, two storey and first floor side extensions to a semi detached, linked detached or end terrace property, should incorporate the following principles:

- The extension should not exceed more than 50% of the width of the frontage of the original dwelling.
- A minimum of 800mm shall be retained between the sidewall of the extension and the inside of the plot boundary to allow for access to the rear for bin and cycle storage. (see diagram 9)
- A minimum gap of 800mm shall be retained between the sidewall of the first floor and the plot boundary. (see diagram 9)
- The extension shall be set back a minimum of 1 metre from the main front elevation of the existing dwelling. (see diagram 10)
- The roof of the extension shall have a lower ridge height, than the existing house. (see diagram 10)

Diagram 9: Good side extension with 800mm retained between the sidewall of an extension and the plot boundary.

- A minimum of two off road car parking spaces shall be provided.

Other considerations

5.2 This policy is designed to prevent extensions at the side of detached or semi detached houses from joining up with neighbouring houses to create a continuous terrace effect. Whilst there is nothing wrong with terraced housing as such, the aim of the policy is to protect street scene and the amenities of areas that were originally designed and laid out as detached or semi detached developments. Such areas can provide attractive views between houses to trees and the scene beyond, and they permit the penetration of sunlight and daylight into the street and into gardens and rooms opposite the gaps. Closure of these important gaps between dwellings can alter the character of a residential street leading to a reduction in the amenity enjoyed by residents and passers by. (see diagram 11)

5.3 For detached properties the lower ridge height and first

Diagram 10: Good side extension – set back 1m and with lower ridge height

Diagram 11: Inappropriate side extensions which have lead to a 'terracing effect'

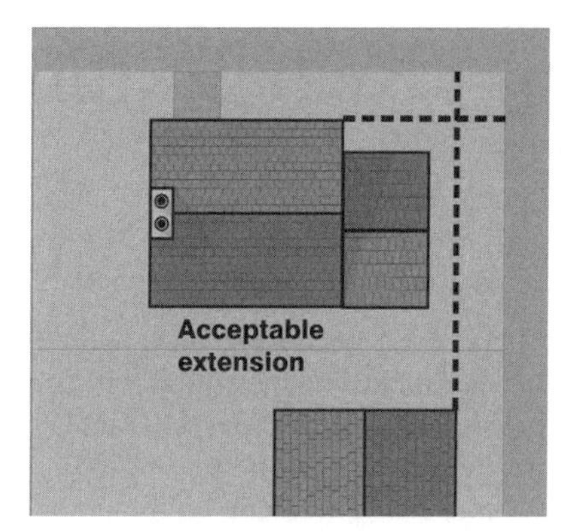

Diagram 12: Appropriate comer extension not projecting beyond either properties front elevation

floor front elevation set back may not be required but this is based on a case-by-case assessment.

5.4 This policy also ensures that the extension is subordinate to the existing dwelling and harmonises with it.

Corner plots

5.5 Even though a corner plot may seem to have more garden space to the front and side, they should remain open, with clear views to be seen when travelling around the corner. The Council will normally expect all extensions on corner plots (single and two storeys) to meet all the following criteria:

- Corner extensions should not project beyond the front elevations of those properties on adjacent roads. (see diagram 12)
- The width of the extension should not be more than half the width of the original frontage of the property.
- The width of the extension should not be more than half the width of the garden / plot between the property and adjacent highway.
- The extension should have a pitched roof to match the design of the main roof.

6 Rear Extensions

6.1 When considering rear extensions the Council will use the 45-degree rule. This will help to assess the impact of any rear extension upon the amenities of the neighbouring properties and to protect them from overshadowing or obstruction, caused by extensions on or close to the boundary.

Single storey rear extensions

6.2 These principles apply to single storey rear extensions:

- An extension will not normally be allowed if it projects more than a 45 degree line from the middle of the nearest affected

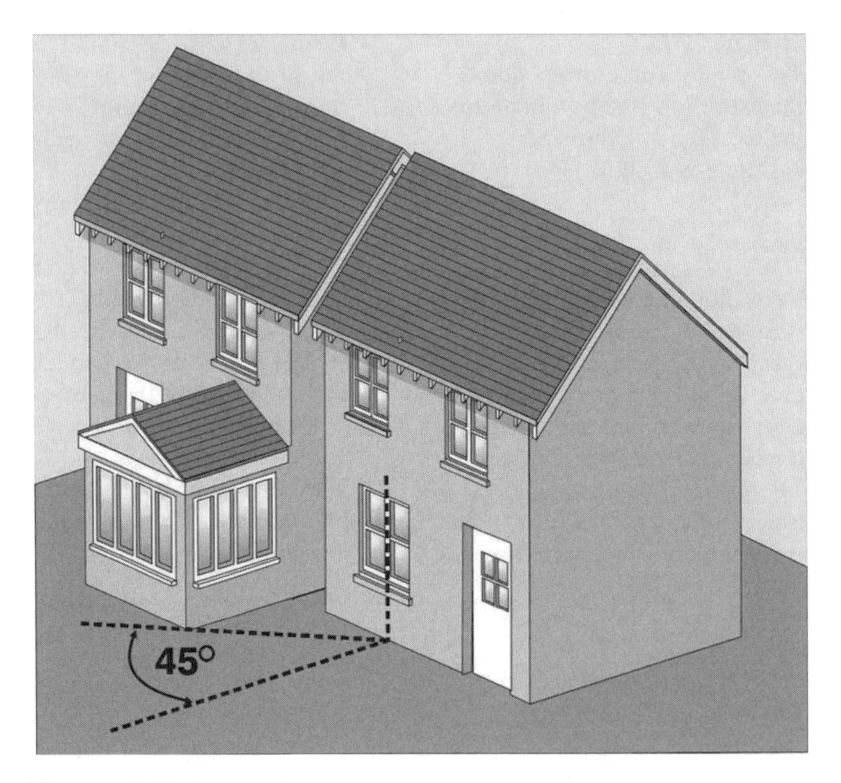

Diagram 13: 45-degree rule

neighbouring window or exceeds a maximum of 4 metres.

- To comply with the 45-degree code, extensions should be designed so as not to cross the 45-degree line from the neighbours nearest habitable room (living, dining, conservatory or bedroom) window. The 45-degree line shall be drawn in the horizontal plane, and taken from the middle of the neighbour's window. The line will show the maximum width and / or depth that a proposed extension can build up to avoiding obstruction from light or views. (see diagram 13)
- The council when assessing single storey rear extension will consider the impact on the neighbouring property and take into account differences in land levels.
- The council will also take into consideration the height of a proposed extension when assessing an application.

Two storey rear extensions

6.3 The following principles apply to two storey rear extensions:

- Two storey extensions along shared boundaries shall not project at first floor level by more than 2 metres.
- In any other case, the following sizes shall be applied:

Distance between extension and adjoining property	Maximum projection at first floor level
1m	2.5m
2m	3m
3m or more	4m

- Where properties have a staggered building line and a neighbouring property is set forward in the plot, the maximum projection will be measured from the rear building line of that neighbours property.

7 Dormer Extensions

7.1 Wherever possible dormer windows should be restricted to the rear of the dwelling in order to preserve the character of the street scene. This may not be so important where front dormers are already a common feature of other buildings in the street.

7.2 Side dormers will not normally permitted where they allow overlooking or adversely affect the streetscene.

7.3 Where dormers are on the front or rear elevation of the dwelling or readily visible from public space, their scale and design are particularly important and the following criteria will apply:

- They should not normally exceed more than one third of the width of the roof.
- They should not project above the ridge of the roof (see diagram 14).
- Dormers which wrap around the side ridges of a hipped roof are not acceptable.
- The face of a dormer should be set back by a minimum of 1 metre behind the main wall.
- A dormer should not extend to the full width of the roof, but should be set in from the side/party walls. Two smaller dormers may be better than one large one (see diagram 15).

Diagram 14: Examples of dormer window extensions

Diagram 15: Examples of dormer window extensions

- Dormer windows should vertically line up with existing windows and match their style and proportions.
- Flat dormer roofs are not acceptable unless considered appropriate to the particular building or the street scene.
- Dormer cheeks should normally be clad in materials to match the existing roof.

8 Parking & Garage Space

8.1 Extensions will not normally be allowed if they have the potential to reduce off-road parking. In most circumstances a minimum of two off-road parking spaces should be provided. The size of a parking space should be a minimum of 2.4 metres × 5 metres.

8.2 It is preferable that at least one parking/garage space is provided behind the building line, and that the driveway can accommodate at least one vehicle length of 5 metres. This should not include any service verge or footpath. Where a garage has been provided adequate space shall be given to allow for a parked car and for a garage door to be opened (see diagram 16). Where space is restricted the use of roller or sectional garage doors which require less space may be more appropriate than 'up and over' doors.

8.3 Where the extension or alteration will create a 5-bedroom property three off road parking spaces will be required unless there are other material considerations.

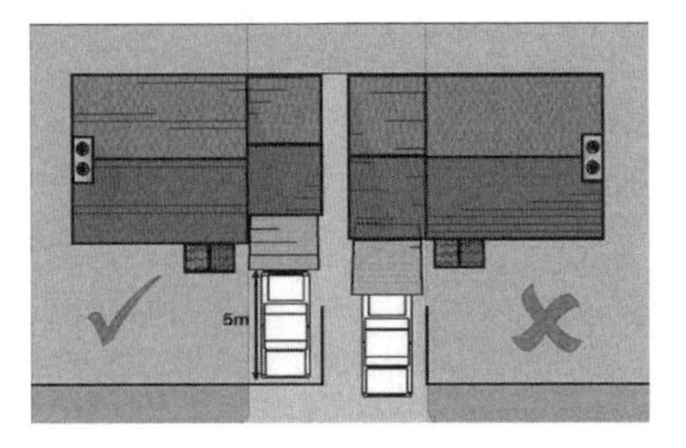

Diagram 16: Examples of parking spaces

8.4 Consideration of parking arrangements is particularly important if you are thinking of either converting an integrated or attached garage to living accommodation or building on or over an existing garage or driveway. Extensions which prevent the parking of cars within the curtilage of the dwelling will not be acceptable.

8.5 An extension should not be constructed in a position where it would interfere with an adequate standard of visibility for road users to the detriment of highway safety.

8.6 Property boundaries, including extensions should be 300mm clear of the highway boundary (including footpath and service verge) so that foundations do not interfere with service apparatus.

9 Garden Space

9.1 Enough private garden space should be left after any extensions have been built to accommodate various leisure pursuits, to ensure that enough space is kept between neighbouring houses to avoid a cramped overcrowded feel and to prevent overlooking between windows.

9.2 The minimum garden area acceptable to the Council is 50 Sq.m of usable garden space, and this should be private enclosed space e.g. rear garden. In most cases it will be necessary to keep a larger area to avoid cramped appearance and to maintain the character of the area. Large trees within gardens may restrict options to extend especially if the trees are protected by a Tree Preservation Order.

9.3 Generally rear extensions will encroach onto garden space to a greater extent than side extensions. If the garden is already quite small (approximately $50m^2$), a rearwards extension may not be advisable.

Trees

9.4 Proposals that would require the felling of protected trees, trees in conservation areas or other trees that contribute significantly to the character of an area or that could endanger its health (for example by severing its roots), are very rarely considered to be acceptable. Instead, alternative methods of providing additional accommodation should be explored.

9.5 Extensions will not normally be allowed where the extension will be overshadowed by surrounding tree, as this could lead to pressure to remove these trees.

9.6 Proposals that would result in the felling of trees or would extend within the canopy of such trees or in close proximity must be accurately shown on the submitted plans, and include the crown spread.

10 Other Considerations & Information

OTHER CONSIDERATIONS

Green Belt

10.1 Extensions should not result in disproportionate additions over and above the size of the original building. The interpretation of this policy will vary according to the character of the property, but as a general guide, extensions, which increase the volume of the original house by more than about one third, are unlikely to be acceptable.

10.2 Special regard should be given to matters of siting, height, scale, design and use of materials in order to maintain the openness and visual integrity of the Green Belt.

10.3 Planning applications for extensions should be accompanied by drawings that demonstrate the size/volume of the original building. Very special circumstances will have to exist to justify any exception to the strict control of development in the Green belt. It is the responsibility of the applicant to provide this justification.

Listed Buildings and Conservation Areas

10.4 Extensions to listed buildings and/or within Conservation Areas are likely to be particularly sensitive. In particular, the design standards applied may be stricter than those previously outlined in this policy document. Most works to listed buildings will require Listed Building consent, even if they do not require planning permission.

Protected Species

10.5 Species such as bats, which use roof spaces as roost or hibernation sites, and birds which nest under the eaves of buildings are protected from harm by law. Applications for development that involve alterations to existing roof spaces, listed buildings, pre-1939 houses, barns or other traditional buildings and, any work involving disturbance to trees or hedges may have an impact upon protected species.

10.6 If the presence of bats or birds is suspected then an application may need to include a survey report, together with details of mitigation measures to safeguard the protected a species from the adverse effects of the development.

10.7 The Council may impose planning conditions or obligations on planning permissions to ensure that these measures are implemented. Such measures may include, for example, avoiding carrying out any work during the bird breeding season, or the inclusion of artificial nest boxes as part of the development.

10.8 The Council may refuse permission for developments where inadequate survey and mitigation details are included with an application.

10.9 For further information please visit the Natural England website: www.naturalengland.org.uk

Flood Risk

10.10 The Environment Agency recommends that in areas at risk of flooding, consideration be given to the incorporation into the design and construction of the development of flood proofing measures. Additional guidance can be found on the Environment Agency website: www.environment-agency.gov.uk

10.11 It should also be noted under the terms of the Water Resources Act 1991 and the Land Drainage Byelaws, the prior written consent of the Agency is required for any proposed works or structures either affecting or within the 8 metres of the tidal or fluvial flood defences.

OTHER INFORMATION

Building Regulations

10.12 In addition to the need for planning permission, house extensions may also require approval under the Building Regulations. These regulations are designed to ensure appropriate standards of design and construction are employed. Approval under these regulations is a separate issue and for further advice, please contact the Council's Building Control Division on 0151 471 7360.

The Party Wall Act

10.13 The Act provides a framework for preventing and resolving disputes in relation to party walls, boundary walls and excavations near neighbouring buildings. It does not resolve boundary disputes but is intended to manage the process of work to or up to the party boundary and includes reference to the right of access to neighbouring properties to carry out works. An explanatory booklet is available should further details be required.

Neighbours

10.14 The Council always recommends that you consult neighbours affected by the proposals before submitting the plans to the council. This may facilitate minor amendments and resolve any unknown issues. In any event, the council will formerly notify neighbouring properties inviting comments, and a period of 21 days is allowed for such comments to be made in writing to the council.

Utilities Infrastructure

10.15 The Council would recommend that any applicant check with United Utilities for the presence of underground utility services, which may restrict where extensions may be built.

11 The Planning Application

Do you need to apply for planning permission?

11.1 If you live in a house, you can make certain types of minor changes to your home without needing to apply for planning permission. These rights, are called Permitted Development Rights. However, Permitted Development Rights will not apply if your property is Listed or in the grounds of a Listed building; in a Conservation Area; and/or has had the Permitted Development Rights removed.

11.2 If you are in any doubt as to whether you require planning permission please write in with details of your proposal including sketches and sizes to: Planning & Policy Division, Environmental and Regulatory Services, Rutland house, Halton, Lea, Runcorn, Cheshire, WA7 2GW.

11.3 Further information in relation to planning permissions and Permitted Development Rights can be found in the Department for Local Government and Communities (DCLG) document 'Planning – A Guide for Householders – What you need to know about the planning system' or on the Planning Portal website at www.planningportal.gov.uk or on the Council's website at www.halton.gov.uk

11.4 It should also be noted that house extensions may require building regulations approval regardless of whether or not they need planning permission. More advice on the building regulations may be obtained from the Council's Building Control Division, contact details can be found in the Appendix.

If you build something, which needs planning permission, without obtaining it first, you may be forced to put things right at a later date. This may prove troublesome and costly. You might even need to remove an unauthorised building!

What to submit with your application

11.5 When you are preparing your application, you should

include the following information:

- 3 copies of the completed planning application forms. Blank copies are available online or from Halton Direct Link receptions. You can also apply online at www.planningportal.go.uk
- 3 copies of the location plan, (OS based) showing your property in relation to neighbouring properties and its position in the street, with numbers of nearby houses clearly indicated, to scale of not less than 1:1250. You should outline the boundaries of your property (including land to the front, rear and sides) in red with any other adjoining land in your ownership outlined in blue.
- 3 copies of the plans and elevations of the house as existing
- 3 copies of the plans and elevations showing the extension proposal, to a scale of 1:100 or 1:50.
- A block plan to a scale of not less than 1:500 showing the distances from the extension to your plot boundaries and showing any other features such as trees, outbuildings.

The block plan should show the relationship of the extensions to neighbouring windows. It should also show the means of access and parking arrangements. Note the block plan should be based upon accurate survey measurements. On sloping sites, plans showing proposed levels may be required. (Diagram 15 provides an example of a block plan)

Please note all plans and elevation drawings should be produced in a metric scale.

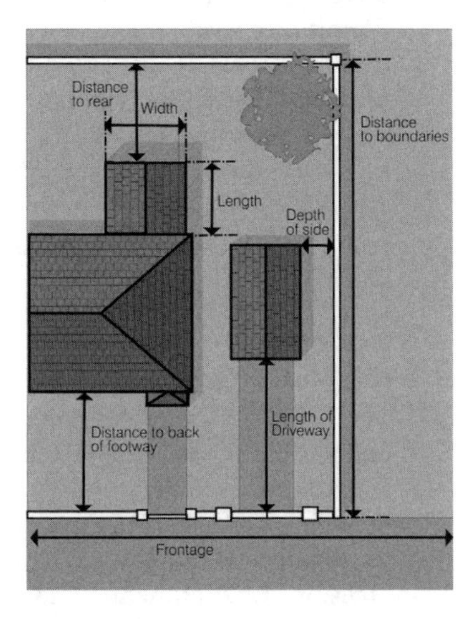

Diagram 17: A typical block plan for a rear extension and garage is shown on the following page.

- A covering letter with any other relevant information in support of you application.
- I copy of the correct Certificate of Ownership, signed and dated.

If you are not the sole owner of all the land to which the application relates the owner/s must be told about your application, this is done by serving Notice No. 1 on each owner.

NOTE: All applications requiring planning permission must be accompanied with the appropriate fee, which must be submitted with the application forms and plans.

Unless the fee is correct, and the forms and certificate are correctly filled in, including accurate plans, the application cannot be registered and will be returned.

11.6 Please note that the advice contained in this document is not binding in every case, so there may be occasions where special site characteristics warrant a relaxation. Officers will always be pleased to advise.

12 Policy Background

12.1 This SPD has been produced to ensure that through its function as a Local Planning Authority, the Council complies with national and regional guidance and advice and contributes, wherever possible, to meeting the priorities of the community its serves.

National Policy

12.2 *Planning Policy Statement I (PPSI): Creating Sustainable Communities*, states that 'good design ensures attractive, usable, durable and adaptable places and is a key element in achieving sustainable development. Good design is indivisible from good planning.'

12.3 Planning Policy Statement 3 (PPS3): Housing, promotes the creation of high quality housing that is well designed and built to a high standard. It highlights the need for places, streets and spaces which meet the needs of people, which are attractive, have their own distinctive identity, and positively improve local character. It also promotes the use of designs and layouts that are inclusive, safe, take account of public health, crime prevention and community safety, ensure adequate natural surveillance and make space for water where there is flood risk.

12.4 'Better Places to Live: A Companion Guide to PPG3: By Design' published by the DETR in 2000 provides specific urban design advice to help deliver the objectives of PPG3. This companion guide considers the principles of urban design and the features of urban form, together with advice on the design and layout of successful residential housing developments, such as understanding character, privacy, orientation and safety.

12.5 Additional good practice guidelines include the Department for Local Government and Communities (DCLG) document 'Planning – A Guide for Householders – What you need to know about the planning system' (2006). Advice contained in this document state that a well-designed building or extension is likely to be much more attractive to you and to your neighbours and it is also likely to add value to your house when you sell it. The guidance for householders specifically suggests that extensions often look better if they use the same materials and are in a similar style to the existing buildings and in some instances the Council's design guides or advisory leaflets may help you or you may wish to consider using a suitably qualified, skilled and experienced designer.

Regional Policy

12.6 One of the core principles of Regional Planning for the North West (RPG13), which is now by virtue of the Planning and Compulsory Purchase Act (2004) the Regional Spatial Strategy (RSS), is good design. Policy DP3 states that 'new development must demonstrate good design quality and respect for its setting'. It goes on to state that local authorities should set out guidance that ensures more innovative design to create a high-quality living and working environment, especially in housing terms, which incorporates: more efficient use of energy and materials; more eco-friendly and adaptable buildings; sustainable drainage systems; community safety and 'designing out' of crime; and appropriate parking provision and best practice in the application of highway standards.

12.7 Policy DPI of the Draft RSS (2006) states that all proposals and schemes must demonstrate excellent design quality, sustainable construction, efficiency in resource use and respect for their physical and natural setting.

Local Policy

12.8 The Halton Unitary Development Plan (UDP), which was adopted in April 2005, contains a number of strategic aims and objectives. These are set out in Part I of the UDP. In relation to environmental quality, these include creating a safe and healthy Halton, and ensuring that future development is of a quality of design that enhances the built environment and encourages the use of energy efficient design. At the centre of these strategic aims and objectives is the desire of the Council to create sustainable places that all people will want to live and work within.

12.9 Part 2 of the UDP contains policies that seek to implement the broad aims and objectives contained within Part I of the UDP.

The proposed SPD is intended to support Policy H6, which states that proposals for house extensions will be permitted where:

a the proposal would not unacceptably alter the appearance or character of the original dwelling but relate closely to it and harmonise with it in terms of their scale, proportions, materials and appearance;
b the proposal would not create dangerous highway conditions by obstructing visibility for pedestrians or drivers of motor vehicles; and
c Reasonable private garden space is provided for use by the residents of the extended property

12.10 However, other policies within the UDP may also be relevant to some developments so this SPD. Therefore, this SPD should be read in conjunction with all the relevant policies of the Development Plan.

12.11 The intended SPD will be produced to contribute to the priorities, principles, objectives and targets of the Halton Community Strategy (2006). This strategy coordinates the resources of the local public, private and voluntary organisations towards common purposes.

12.12 Two of the main priorities set out in this strategy cover issues that are expected to be raised in the proposed SPD, within the priority to Halton's urban renewal one of the objectives to support and sustain thriving neighbourhoods and open spaces that meet people's expectations and add to their enjoyment of life. Within the priority to a Safer Halton one of the objectives is to create and sustain better neighbourhoods that are well designed, well built, well maintained and valued by the people who live in them, reflecting the priorities of residents to improve public perceptions and attractiveness.

12.13 Halton Borough Council is signed up and committed to contributing to achieving the priorities of the Community Strategy. The Council's priorities are set out in the Corporate Plan. This plan also has five priorities, including 'safe and attractive neighbourhoods' and 'promoting urban renewal'.

12.14 The intended SPD is being produced to help meet this target and others set out within the Council's Corporate Plan. The priorities in the Community Strategy and the Corporate Plan are based on the priorities set by the people of Halton. These were identified through community involvement via area panels, focus groups, and a telephone questionnaire. The Corporate Plan was based on the same community involvement and statistical information compiled for the State Of Borough Report, 2005.

Appendix I Contacts and Useful Information

General information

To access a downloadable copy of the Planning Policy Guidance notes or Planning Policy Statements detailed in Section 2, or for further general planning information visit the Department of Communities and Local Government website at www.communities.gov.uk or for a hard copy contact the Department of Communities and Local Government by phone on 0870 1226 236.

To access a downloadable copy of 'By Design, Urban Design in the planning system: Towards Better Practice' and 'Safer Places', documents relating to urban renewal, urban design and creating sustainable communities, and general planning information visit. The Department of Communities and Local Government website at www.communities.gov.uk.

For information relating to urban design there are several documents available. Design at a Glance: A quick reference to national design policy, Design Review and The Value of Good Design can be downloaded free of charge from the CABE website at http://www.cabe.org.uk/publications/ and The Urban Design Compendium produced by English Partnership and the Housing Corporation can be ordered online free of charge from English Partnerships at www.englishpartnerships.co.uk Urban Design Guidance: urban design frameworks, development briefs and masterplans, produced by the Urban Design Group, and From Design Policy to Design Quality, produced by the RTPI, can be purchased from Thomas Telford Ltd.

Further information on the Secured By Design initiative, including details relating to the standards required for a development to receive Secured By Design accreditation may be found at www.securedbydesign.com

For information regarding any development affecting a historic building or conservation area 'Building In Context' will be able to provide advice. It is available from English Heritage and the CABE and can be downloaded free of charge from http://www.cabe.org.uk or for a hard copy contact English Heritage at: Customer Services Department, PO Box 569, Swindon, Wiltshire, SN2 2YP, Tel: 0870 333 1181, Fax: 01793 414 926

You can find out about the planning system and how it works at www.planningportal.gov.uk

Local information

For advice relating to submitting a planning application, for pre-application discussion or to purchase a copy of this SPD or any other SPD contact:

Planning & Policy Division

Environmental & Regulatory Services
Halton Borough Council
Rutland House
Halton Lea
Runcorn
WA7 2GW

Tel: 0151 424 2061
Fax: 0151 471 7314
Email: dev.control@halton.gov.uk or forward.planning@halton.gov.uk
Website: www.halton.gov.uk/development control or www.halton.gov.uk/forwardplanning

If further highways or transport information is required, please contact the:

Highways Division

Environmental & Regulatory Services
Halton Borough Council
Rutland House
Halton Lea
Runcorn
WA7 2GW

Tel: 0151 424 2061
Fax: 0151 471 7521

If further information is required relating to accessibility or building control please contact:

Building Control Division

Environmental & Regulatory Services
Halton Borough Council
Rutland House
Halton Lea
Runcorn
WA7 2GW

Tel: 0151 424 2061
Fax: 0151 471 7314
Website: www.halton.gov.uk/building control

If further information is required in relation to trees in development, please contact:

Landscape Division

John White (Trees & Woodlands Officer)
Environmental & Regulatory Services
Landscape Services Department
Picow Farm Depot
Picow Farm Road
Runcorn
WA7 4UB

Tel: 0151 424 2061
Website: www.halton.gov.uk

Index

Page numbers in **bold** refer to figures.